ALSO BY IAN AYRES

Responsive Regulation:
Transcending the Deregulation Debate

Studies in Contract Law

Voting with Dollars:
A New Paradigm for Campaign Finance

Pervasive Prejudice?
Unconventional Evidence of Race and Gender Discrimination

Why Not?
How to Use Everyday Ingenuity to Solve Problems Big and Small

Straightforward:
How to Mobilize Heterosexual Support for Gay Rights

Insincere Promises:
The Law of Misrepresented Intent

Optional Law:
Real Options in the Structure of Legal Entitlements

Super Crunchers:
Why Thinking-by-Numbers Is the New Way to Be Smart

Lifecycle Investing:
A New, Safe, and Audacious Way to Improve the
Performance of Your Retirement Portfolio

CARROTS
and STICKS

CARROTS and STICKS

UNLOCK THE POWER OF
INCENTIVES TO GET THINGS DONE

———

IAN AYRES

BANTAM BOOKS
NEW YORK

Published in the United States by Bantam Books,
an imprint of The Random House Publishing Group,
a division of Random House, Inc., New York.

BANTAM BOOKS and the rooster colophon are registered
trademarks of Random House, Inc.

LIBRARY OF CONGRESS CATALOGING-IN-PUBLICATION DATA

Ayres, Ian.
    Carrots and sticks: unlock the power of incentives to get things
done / Ian Ayres.
p.   cm.
    Includes bibliographical references and index.
    ISBN 978-0-553-80763-9
    eBook ISBN 978-0-553-90782-7
       1. Commitment (Psychology)   2. Incentive (Psychology)
3. Motivation (Psychology)   4. Economics—Psychological aspects.
I. Title.
BF619.A97 2010
158—2010   009939dc22

Printed in the United States of America on acid-free paper

www.bantamdell.com

9 8 7 6 5 4 3 2 1

First Edition

Book design by Steve Kennedy

For Dean Karlan,

who asked, "Why not a commitment store?"

Contents

Author's Note

The books I value most are those that provide a central insight that, once recognized, changes the way you view the world and provides you with a tool for action. I like news you can use. I've written this book because its key idea, that finding the right incentives is often the difference between success and failure, has been both the subject of my work and the source of some significant positive changes in my own life. My hope is that once you see how new incentive structures can, for example, help you quit smoking, lose weight, or better manage your employees, you'll begin to recognize other areas of your professional and personal life in which you can craft incentives to reach a desired result.

Introduction

SNEEZE

On January 14, 2008, Alex Moore, an electrical engineer and recent graduate of MIT, did something strange. He promised not to "artificially sneeze" for eight weeks. Even stranger, he backed up his promise by putting $400 at risk. If in any of the next eight weeks he artificially induced himself to sneeze, Alex would lose fifty bucks. And to make the loss even more painful, any money he lost would be donated to an anti-charity, that is, a charity that Alex doesn't support.

When I learned of Alex's no-sneeze promise, my first reaction was to think it was some kind of weird MIT practical joke. (I went to MIT, and it's the kind of place where students measure the Mass Ave bridge in "smoots.") But it turns out that Alex was serious. "When I was in tenth grade," he said, "my English teacher told us that during the Enlightenment, all the great thinkers would roll tobacco into thin tendrils, then tickle the insides of their noses to make themselves sneeze. That night, I was curious to see if it really worked, so I tried with a toothpick. Instant success! And I loved it. You know how a good sneeze just makes you feel invigorated? When I'd read books, I'd sneeze. When I'd do homework, I'd sneeze. I'd make myself sneeze over and over until I'd sound like I had allergies or a head cold for a couple days."

Over the years, Alex tried several times to quit. He made New Year's resolutions, but to no avail. "I'd go a few days without sneezing," he told me, "and then I'd start again. I've known I should stop basically since I started, but every time I try, it's hard, and I'd forget once or twice, and then all bets are off."

So on that fateful day in January, Alex entered into a legally binding commitment contract at a new Internet site, stickK.com, in a serious effort to break the cycle of repeated attempts and failure. He risked forfeiting $400 to the National Center for Public Policy Research (which he described as an "anti-environmental" charity) because he "figured that would be even more incentive than the money." The threat of losing $400 kept him on the straight and narrow.

And the good news is that it worked. "After probably thirty attempts to stop completely, I have gone two months without a single sneeze that nature didn't intend," Alex enthused. "I went for a run this afternoon, and my lungs were completely clear." What's more, Alex got his $400 back. He put a lot of money at risk, but at the end of the day the commitment contract didn't cost him anything.

Few people use toothpicks to induce the momentary loss of control of a sneeze. It's easy to think of Alex as cut from a different kind of cloth than the rest of us. But there's a little bit of Alex in all of us. I have never felt the impulse to make myself sneeze, but I have wasted huge amounts of time playing Minesweeper. Almost all of us engage in some activities that we know aren't really good for us. We procrastinate. We make resolutions that we don't keep. So instead of seeing Alex as someone different, one challenge of this book is to see how we're similar.

More importantly, Alex's solution to his problem, a commitment contract, may be the key to helping people lead happier lives. Commitment contracts provide a simple, powerful, and unifying solution to a host of behavioral problems, and behavioral economics is showing that they really do work. Commitment contracts are promises backed by contingent rewards or punishments. With a commitment contract, if you promise to exercise three times a week, you had better do it or you'll be hit by some kind of penalty (or lose out on some kind of reward). Isn't this just an incentive? Yep. People have been using various forms of incentive contracts forever—from Odysseus to Curt Shilling, who in 2007 entered into a $2 million weight-loss incentive contract with the Boston Red Sox.

But this book is centrally about how to craft commitments that will work best for you. The new learning of behavioral economics has a lot to say about how best to tailor commitments to make them more effective and virtually free. There are dozens of different choices for how to structure contractual commitments. For example, was it crucial that Alex put so much money at stake or that he designated an anti-charity to receive that money if he lost?

I'm not a neutral bystander on these questions. I care passionately about the benefits of commitment contracts and have already put that passion into action. Together with Yale economist Dean Karlan and Yale business student Jordan Goldberg, I founded the service that Alex used to kick his sneezing habit.

stickK is not just about sneezing. It's a commitment store that you can use to help yourself stick to just about any goal. You choose the form of accountability that works for you. You can choose to referee your own contract ("on your honor") or you can designate any other person in the world to be your ref. We also provide a nagging service; or, if you prefer, you can put your reputation at stake and we'll tell your designated friends and family members whether or not you succeeded. Most important, we let you put money at risk that will be forfeited if you fail to stick to your goal. You can even choose who gets the money if you don't succeed—we provide a list of anti-charities.

I put my money where my mouth was by sinking a bunch of my own savings into stickK. But I'm committed to the value of stickK as more than just a business idea. Like the guy in the Hair Club for Men ads, I'm not just a founder of the company; I'm one of its clients. At the beginning of 2007, I was massively out of shape. I had stopped exercising the previous fall and my weight had ballooned to 205 pounds. I plugged my height and weight into a BMI calculator and was aghast to learn that this skinny kid from Kansas City who, in high school, couldn't put on a pound to save his life was now officially overweight (with a BMI greater than 25).

I resolved to get back to 180. To be honest, it had been easy for me to lose weight in the past. But I've never been able to keep it off. My weight would yo-yo. Four or five different times over the last decade, I've gone on a crash diet and lost a bunch of weight, only to regain it before the year was out.

So in 2007, I did something different. I backed up my weight-loss resolution by putting $500 at risk each week for the entire year. To start off, I had to lose a pound a week until my weight dropped below 185. And then I had to keep my weight below 185 for the rest of the year. (I really wanted to weigh 180, but I figured I should give myself a safety cushion of five pounds to allow for normal fluctuations.) All in all, I put $26,000 at risk. And Dean Karlan was ready to pounce if I failed. Dean also feels passionately about commitment contracts. He's refereed commitment contracts in the past and had taken thousands of dollars in stakes from one of his friends. He had just the reputation I wanted.

In figure 1 we have a picture that's worth a thousand words.

Figure 1. Tracking Ian's Commitment Contract

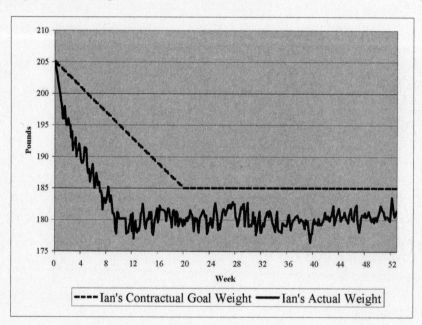

You can see that my weight went from 205 down to 180 in about ten weeks. But what is truly remarkable is how steady my weight remained after I first hit 180. And it's remained remarkably flat to this day (I weighed 181.2 pounds this morning).

Before stickK, I knew how to eat a 160-pound diet (which I could sustain for short periods when I wanted to lose weight quickly) and I knew how to eat a 200-pound diet, but a commitment contract has taught me how to eat a diet that will support my 180-pound target weight. Weight Watchers costs about $500 a year, and the average person loses about six or seven pounds by the end of the year. I put $500 at risk each week, but in equilibrium, I've lost twenty-five pounds (12 percent of my body weight), and so far it has cost me nothing. I decided to write this book because I honestly believe that commitment contracts can help people improve their lives as they have helped me.

AGAINST CHEERLEADING

This book, however, is not an extended advertisement for stickK or for the value of commitment contracts. We will explore not only how best to pick the right commitment tool but also when it's best to keep the tool in the box. If we really want to learn about when commitments work, we also need to look at examples of when they fail.

For example, back in 2007, Simon Usborne, a reporter for the British newspaper the *Independent,* asked if he could interview me for an article. But when the appointed time came, he never called. He emailed later to reschedule, then blew me off again. When he asked to reschedule a third time, I felt a bit like Charlie Brown being asked by Lucy to trust her and go ahead and kick the football. But to be honest, I had a book (*Super Crunchers*) that I was trying to promote, so I gamely agreed. And once again I was left standing at the altar. You'd think this would be the end, but Simon (and this is his real name) emailed me again:

> Ian,
> I feel very bad for putting you off once and then mixing up the times today.
>
> But things are now so busy that an interview would be rushed, and I'd much rather give it my full attention when it's less frantic. That is, of course, if you aren't fed up with me and we can find a new time that suits you.
>
> I'm away tomorrow and Friday. How does Tuesday 10am your time, 3pm mine, sound?

Any sane person would have declined his kind offer. But I took a different tack. I tried to use a commitment contract to get him to change his low-down ways. So I emailed Simon back:

Tuesday, Sept. 4 at 10 a.m. Eastern is fine by me.

I'm not fed up. But I would appreciate trying to get it done then.

How about promising to give 50 pounds to the Multiple Sclerosis Society (my mom died of MS) if my office phone doesn't ring by 10:05?

I added in the true detail about my mom to make clear that I wasn't kidding. Commitment contracts are valuable not only to help people like Alex change their own behavior; they're also valuable because they can help us change other people's behavior. Some people tell me that stickK is a cool site but they don't need it because they don't have trouble following through on their goals. Another theme of this book, however, is to show that we all need commitments, even if not for ourselves, because we all encounter people like Simon as we walk through life. Commitment contracts are useful not just because there is a little bit of Alex in many of us but also because there is a little bit of Simon in others.

You're probably wondering how Simon responded to my proposal. He fatefully replied:

Your phone will ring!
Thanks Ian, Simon

But the dirty dog didn't call at the appointed time, and he didn't pay (even after I helpfully sent him a link to the Multiple Sclerosis Society, U.K.). Here's an example where a commitment contract—or at least an attempt at a commitment contract—failed. As a contract scholar, I know there's a real chance that an English court would find Simon's ambiguous reply to my offer sufficient to create a binding agreement. But what's important is what Simon thought. For a commitment contract to be effective, Simon had to believe that he was bound. I think it's likely that Simon tried to thread the needle—to convince me to agree once again to be interviewed without actually taking on the obligation to pay if he failed. (Simon didn't respond to my emails asking to interview him for this book.) I could have written back, demanding clearer words of acceptance: "So you are agreeing to pay the MS Society if you don't call?" But I didn't. By failing to pin him down, I failed to create an effective commitment. My frustration with people like Simon is one of the things that motivated me to found a website where people can quickly post a charitable bond to back up their promises. If stickK had existed in 2007, I could have asked that Simon create an enforcement bond payable to MS and make me the referee of his commitment.

LOST THE MAGIC

However, the problem of commitments is not just that they are sometimes unclear or ambiguously enforceable. Indeed, Alex's initial success with a commitment contract was followed by commitment failure. "I actually took out another contract on stickK," he told me, "because I was blown away that it worked so well. I am one of those people that have trouble getting to work on time in the mornings—I think it is common among MIT Course 6 graduates. So I took out a contract that said that I would get to work by nine-thirty A.M." As with his initial sneezing contract, he put $50 at risk each week that, if he failed, would go to an anti-charity (this time, the George W. Bush Presidential Library). Unlike in my email exchange with Simon, there was no legal ambiguity. Alex even upped the ante by making his girlfriend the contract's referee. Just like Dean Karlan, she could independently verify whether he kept his commitment.

But the results were not as successful. "The contract worked for about a week," Alex said. "Then I got stuck in traffic one morning, and through no fault of my own I ended up missing the time. I really didn't know what to do, and I didn't want to send money away to causes that I think are evil for something that I didn't really have any control over, so I let that one slide.

"It was kind of a slippery-slope thing," he continued. "One day the next week I just didn't get out of bed and so I got into work and I thought, 'Oh crap!' I got the notice and the email that I had to do the report, but I talked with my girlfriend about it and I ended up sending the same amount of money to a charity and lying on the stickK contract. So after that I kind of lost the magic."

This book is a search for Alex's lost magic. Why is it that these seemingly similar contracts produced very different results? Alex succeeded on the harder task of overthrowing an ingrained lifetime habit, but couldn't bring himself to get out of bed. At the very least, we know that commitment contracts are not a panacea. But I hope to convince you that by attending to their structure, we can better stick to our commitments. Small differences in detail matter. It is foolhardy to enter into any old commitment without thinking about close to a dozen different dimensions of design. In the following pages, I will introduce you to a wealth of field experiments by behavioral psychologists and economists that are beginning to show us how best to design commitments to maximize our chance of success.

At the end of the day, commitment contracts have helped Alex improve his life. He still forgets himself and induces a good sneeze from time to time. But now he's down to doing it only about once every two weeks. "It's not nearly as frequent," he said. "I'll do it once and I will kick myself, and then another two weeks will go by and then I will be thinking about something and I will do it again." It's still a struggle for him, but he's made progress. And that's a reasonable jumping-off point for this book. Commitment contracts are not magic pills that automatically make everything better. But they can help. We'll learn more about what we can do to make them work better (and when we'd be better off forgoing them altogether).

CARROTS
and STICKS

1

Thaler's Apples

The behavioral revolution in economics began in 1981 when Richard Thaler published a seven-page letter in a somewhat obscure economics journal. Richard is now a stocky sixty-three-year-old with unruly gray hair who looks more like a bartender than one of the world's leading economists. But back then, a thirty-five-year-old Richard posed a pretty simple choice about apples.

Which would you prefer:
  (A) One apple in one year or
  (B) Two apples in one year plus one day?

This is a strange hypothetical—why would you have to wait a year to receive an apple? But choosing is not very difficult; most people would choose to wait an extra day to double the size of their gift.

Thaler went on, however, to pose a second apple choice.

Which would you prefer:
  (C) One apple today or
  (D) Two apples tomorrow?

What's interesting is that many people give a different, seemingly inconsistent answer to this second question. Many of the same people who are patient when asked to consider this choice a year in advance turn around and become impatient when the choice has immediate consequences—

they prefer C over D. When it comes to apples, Adam and Eve aren't alone in being impatient when presented with an immediate temptation.

The inconsistency in these answers puzzled Thaler. Richard has an incredible eye for anomalies. While many economists ignore or paper over deviations from rationality, Thaler is drawn to them. He's made a career of trying to understand them, and he even writes down algebraic formulas to capture their essence. Why would he mow his own lawn to save $15 but wouldn't be willing to cut his neighbor's lawn even for $25? Why would his decision about whether to drive through a snowstorm to see a basketball game turn on whether he paid for or was given the ticket? Thaler's obsession with the failure of traditional economics may end up earning him the Nobel Prize. By integrating psychology into economic theory, Thaler and a cadre of other behavioral economists have remade the landscape of economic thinking.

Today the defenders of the faith in economic rationality are still dominant in economic departments across the country, but they are increasingly acknowledging the power of behavioral predictions. While economists have traditionally focused on information and incentives as the core levers of human behavior, Thaler often thinks about what he calls "choice architecture." Simply and slightly reframing the context in which decisions are made can have big effects on people's behavior and happiness. Thaler is even willing to reengineer the rules of golf.

Cade Massey, a Yale business professor, played golf with Thaler soon after Thaler had taken up the game. "Dick suggested that we play with a different scoring rule," he recalls. "You'd get one point for every hole that you parred and nothing for every other hole, regardless of how badly you did. The winner was the person with the most points. This simple rule change really improved my enjoyment. The problem with traditional scoring for a golfer like me is that I could screw up a decent round with a bad score on a single hole." Richard's method allowed Cade to focus more on his successes than his failures. The story also shows how Thaler is constantly trying to use what he knows about how our minds work to increase people's happiness.

The golf purists out there will resist the idea that a scoring change could make golf more enjoyable. Or they might insist that it wouldn't be golf anymore. But Thaler has gone after much larger fish. His impulse to tinker with the rules of the game led him to champion the Save More Tomorrow program. Thaler saw that lots of people were having trouble taking advantage of their companies' 401(k) plans—even when their employers were willing to match their contributions dollar for dollar. It just hurts too much to see your current paycheck go down. But Save More Tomorrow lets people sign up for participation without their current or future paychecks ever declining. Savings contributions are made only out of a portion of future pay increases. These contributors usually reach full participation within four or five years. Over the course of a working life, this is a fairly small delay. But Richard's small change has shown big results. Employees who would never have otherwise invested in a 401(k) plan have been saving for retirement at nearly the same rate as those employees who chose to invest in their 401(k) plans from the start.

The seeds of this idea were planted back with those simple questions about whether you'd be willing to wait an extra day for an extra apple. Thaler's curiosity about the inconsistent answers to the two questions helped spark a revolution in economic thinking. Why were people more impatient when the proffered gifts were closer at hand? Both hypotheticals ask us if we're willing to wait an extra day to double the size of our gift. The only difference is the time perspective.

Now, one way out of this conundrum is to focus on trust. We're taught the old maxim that "a bird in the hand is worth two in the bush," because promised future paydays don't always materialize. The people who prefer an apple today may not trust that they'll actually be given two apples if they wait till tomorrow (whereas they don't think there's much of a difference in the likelihood of being given one or two apples if they have to wait a year).

But Thaler thought there was something else going on.

What was revolutionary about his apple example is that it illustrated the plausibility of what behavioral economists call "time-inconsistent" preferences. Richard was centrally interested in the people who chose both B and C. These people, who preferred two apples in the future but one apple today, flipped their preferences as the delivery date got closer.

Indeed, we can imagine anchoring the dates somewhere in the distant future (Would you prefer an apple on April 4 or two apples on April 5?) and then revisiting the question on a daily basis. Initially, when April is far away, people would choose to double their gift by taking the April 5 option. But because of a time-inconsistent preference, there would come a day where suddenly people would shift to the option of one apple on April 4. This flip in preference is the key to the behavioral revolution in economics.

If we change the hypothetical from apples to crack, it's easy to understand this kind of time inconsistency. A drug-deprived addict may prefer an immediate fix to a double dose tomorrow. When we hear about someone like Alex choosing the immediate small pleasure of induced sneezing to the larger future pleasure of clear lungs, we might be quick to think of him as having an addictive personality or some kind of OCD (obsessive-compulsive disorder). But Thaler's apple example shows that you don't need psychopharmacological dependence to exhibit irrational impatience. Even with regard to something as prosaic and healthful as an apple, many people in the world become impatient as their time horizon shrinks.

In at least some aspects of their lives, many people are addicted to *now*. We are seduced by the immediacy of gratification. The apple choice is replayed repeatedly throughout our lives. My kids put off cleaning their rooms and then waste huge amounts of time looking for a misplaced sweater or birthday invitation. A friend of our family again and again put off buying life insurance and then tragically left his family unprotected. The marginal convenience of delaying a mammogram or other mildly unpleasant task blinds us to the exponential pain just around the bend. I turned fifty almost a year ago, and I still haven't had a colonoscopy.

If asked in late January, most of us would say we would rather spend eight hours working on our taxes on April 1 than twelve hours on the day of the filing deadline, April 15. But as April Fool's day approaches, many of us (foolishly) will reverse course and choose more pain as long as it is even slightly postponed. This is no hypothetical for me. I postpone my visit with H&R Block year after year, even when I expect a fat refund.

Behavioral economists explain these reversals with the esoteric term "hyperbolic discounting." Rational choice theory predicts that the value today of some future gift should lose a fixed proportion of its value for every unit of delay. Rational actors, for example, might think that the present value of an apple decays 1 percent every extra day you have to wait for it. Rational actors would never be impatient if asked either of Thaler's questions because two apples, even with the extra one-day 1 percent discount, would always be better than one apple without this discount. But for decades behavioralists have noticed that humans tend to discount delays hyperbolically—there are big reductions in value for the first small bits of delay and relatively small reductions in value for subsequent increases in delay.

Hyperbolic discounting can explain why people are more likely to switch to the smaller but sooner reward when the reward is close at hand. In the apple example, a hyperbolic discounter might feel that delaying the immediate receipt of an apple by just a day would reduce its value by 70 percent, but delaying the receipt from a year to a year and a day would reduce the apple's present value only from an 80 percent discount to an 81% discount. The hyperbolic discounter will prefer a single apple now to two apples (discounted at 70 percent) a day from now; whereas when confronted with the analogous choice to receive two apples a year in the future, she will hold out for the two apples in part because she so strongly discounts any reward that far in the future. Hyperbolic discounters put extraordinary value in receiving rewards immediately (and in pushing off immediate burdens for even short periods), but then become relatively indifferent about when the reward (or burden) arrives in the future. For them, a reward two years from now is not that much better than a reward five or ten years from now. This is a bit of an overstatement, but a hyperbolic discounter tends to think it's "now or whatever."

Hyperbolic discounters are classic "conflict avoiders." They postpone disciplining the wayward child or addressing marital strife because when confronted with a "pay me now or pay me later" choice, they strongly prefer bearing the later but larger cost. These inconsistent preferences are more than a theoretical possibility. I want to ultimately convince you that many people manifest increasing impatience as the time before the moment of the reward shrinks.

BIRD BRAINS

But let me start by telling you about pigeons—six male White Carneau pigeons, to be exact. Back in 1981, the same year that Thaler published his famous apple study, two Harvard researchers published a laboratory experiment testing the same idea on a very different kind of subject. Over an eleven-month period, George Ainslie and Richard Herrnstein presented the six pigeons with more than 75,000 food choices. Once a day, each pigeon was placed in a special feeding apparatus that had a green and a white typewriter key mounted about five inches apart on a wall. For each feeding test, the two keys would light up and the pigeon had 10 seconds to peck one of the two keys. If the pigeon pecked the green (left) key, a food hopper—after a delay of "D" seconds—would open to let the pigeon eat for 2 seconds. However, if the pigeon pecked the white (right) key, the food hopper—after a delay of "D plus 4" seconds—would open for 4 seconds. For example, when D was equal to 1, the pigeon could choose to have 2 seconds' worth of food if it was willing to wait 1 second, or 4 seconds of access to food if it was willing to wait 5 seconds. Sixty seconds later, both keys would light up again and the pigeon would be confronted with the same choice. This process would repeat for about forty-five tests each day for each pigeon for eleven months (thank God for laboratory assistants).

Notice how closely this experiment parallels Thaler's apple hypothetical. In all 75,000 tests, the pigeons were given the opportunity to eat twice as much (4 seconds of access to food versus 2 seconds) if they were willing to wait a bit longer. The only difference—as with Thaler—was in the base delay, D. For example, in some sessions, the pigeons were asked to choose between waiting 12 seconds and waiting 16 seconds (i.e., D was equal to 12).

The researchers obviously couldn't explain the rules of the game to the pigeons. But by trial and error, the pigeons would learn about the nature of their environment and the relative costs and benefits of the two keys. Usually within just a few sessions, the pigeons would develop fairly stable preferences for one key or the other. The researchers maintained a constant D environment until all the pigeons' responses became stable. To keep the pigeons motivated (i.e., hungry), the researchers maintained the pigeons at 80 percent of their free-feeding body weight.

This bizarrely simple experiment provided strong evidence that you don't need to be a rocket scientist to follow the dictates of hyperbolic discounting. You don't even need to be human.

The graph in figure 2 shows the proportion of feedings across the six avian subjects where pigeons chose patience—that is, where the pigeons, by pecking the white key, were willing to wait 4 seconds to double their opportunity to eat. First, for those who think that literal birdbrains are just too limited to do more than randomly peck in search for food, notice

Figure 2. Preference Reversal in Pigeons

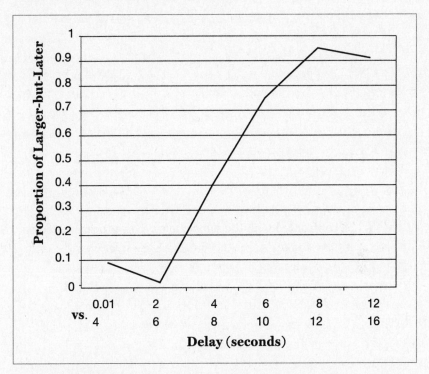

Source: George Ainslie and R. J. Herrnstein, "Preference Reversal and Delayed Reinforcement," *Animal Learning and Behavior* 9 (1981): 478, fig. 3.

how nonrandom the pecking patterns are. The proportion of birds choosing patience when the delay was short (say, 2 seconds versus 6 seconds) was dramatically lower than the proportion when the delay was longer (say, 8 seconds versus 12 seconds). These pecking patterns do not bounce around 50 percent, but show stark variations contingent on delay.

The experiment provides vivid support for the idea that something like hyperbolic discounting can lead to preference reversals. When presented with the chance of almost immediate gratification of 1 or 2 seconds, the pigeons are almost never able to bring themselves to wait a bit longer to double their food. The proportion choosing patience for these conditions is less than 5 percent. This result is especially striking as the pigeons learn that they will have subsequent feeding opportunities every 60 seconds. Looking at just the left-hand side of the graph, it would be easy to conclude that these silly animals are just hardwired for extreme impatience.

But when the base level of delay is increased just a bit, the pigeons behave very differently. When choosing between 8 and 12 seconds, it becomes a no-brainer to wait the extra 4 seconds to double your access to food. Just as predicted by hyperbolic discounting, as the time horizon decreases, pigeons become impatient.

The human mind is much, much more sophisticated than a pigeon's. Most years I teach Corporate Finance, where students learn how to use Excel to exactly calculate the discounted present value of future payoffs. Surely, we would make wiser decisions than pigeons.

But Ainslie turned up the same result when humans were asked about cash. In 1993, he found consistent preference reversals when human subjects were given the choice of a cash gift on Friday or a 25 percent larger gift if they waited until the following Monday. When asked in advance most of the subjects preferred waiting the three extra days. But when asked again on Friday itself, nearly 60 percent of those who had initially been patient changed course and chose the smaller, sooner option.

You can replicate this study with your friends. Flip a coin. If it comes up heads, ask, "Would you rather receive $50 today or $100 in six months?" If it comes up tails, ask, "Would you rather receive $50 in one year or $100 in 1 year plus six months?" You'll quickly find that a larger proportion of people choose smaller and sooner as the time horizon declines.

Hyperbolic discounting can be seen in real-world shifts all around us. Many pregnant women who prefer the delayed benefits of natural childbirth when asked before labor change their minds when push comes to shove. A longitudinal study of pregnant women showed a dramatic shift toward the more immediate reward of anesthesia once they were in active labor. For example, look what happened over time when expectant mothers were asked to respond, using a scale of 0 to 100, to "How important is it to you to deliver your child without anesthesia?" and "How concerned are you about avoiding hard labor pains?"

The relative importance of anesthesia and pain management are shown as the value to avoid anesthesia minus the value to avoid hard labor pains. So if a woman rated the importance of avoiding anesthesia at 70 and the importance of avoiding hard labor pains at 20, she would be recorded as having a net preference for avoiding anesthesia of 50. Overall, we see that when asked one month before labor and during early

Figure 3. Changes in Mothers' Preference During Different Stages of Labor

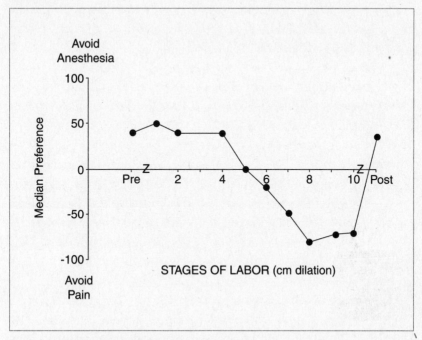

Source: J. J. Christensen-Szalanski, "Discount Functions and the Measurement of Patients' Values: Women's Decisions During Childbirth," *Medical Decision Making* 4 (1984): 52, fig. 4.

labor, the pregnant mothers had a marked preference for the longer-term benefits of avoiding anesthesia. However, at the beginning of active labor (when the woman's cervix was dilated to more than 5 centimeters), the preference shifted markedly toward avoiding pain. Yet when asked at one month postpartum, their preferences shifted back again toward the longer-term benefit of avoiding anesthesia.

Active labor is the quintessential short-horizon moment. To behavioral psychologists, it comes as no surprise that women who genuinely believe the long-term benefits of natural childbirth outweigh the short-term benefits of reduced pain from anesthesia make a different decision when the decision horizon shrinks to almost nothing. Skeptics might respond that women changed their minds during active labor simply because they had new information—that the pain was much worse than they had ever imagined. (I have a friend who was so grateful for the anesthesia that she told me she considered naming her child Epidural.) But this story doesn't explain why their preferences flipped again postpartum. And, more importantly, it doesn't explain why the study found the same pattern in pregnant woman who had previously given birth. Non-first-timers were more likely to stay the course, but they still experienced the double reversal shown by mothers overall.

> *The best laid schemes o' mice an' men*
> *Gang aft agley*
>
> ROBERT BURNS

The truly radical notion is not just that there is a bit of Alex in all of us, but that there is a bit of the pigeon in us as well. The predisposition to hyperbolically discount, according to Cornell behavioral economist (and *New York Times* business columnist) Robert Frank, "is apparently part of the hard-wiring of most animal nervous systems." After reviewing dozens of studies of humans and other animals, Ainslie now claims that the tendency to hyperbolically discount is a "basic function by which all vertebrates devalue delayed events."

But to identify this predisposition is not to say that we are doomed to bear the real costs of irrational impatience. The predisposition is just that—only a tendency—and it varies not only across contexts but across humans. The famed psychologist Walter Mischel studied this variation in impatience—not by asking adults about apples or money but by asking four-year-olds about marshmallows. For six years at the Bing Nursery School, near Stanford University, Mischel and his colleagues asked hundreds of four-year-olds to make a simple choice. Children were escorted individually into an experimental room and were seated at a table on which there was a bell. Each child was shown two plates; one plate had on it one marshmallow, and the other plate had on it two marshmallows. The experimenter told the child that she or he had to go out of the room but that "If you wait until I come back, you can have two marshmallows. If you don't want to wait you can ring the bell and bring me back any time you want to. But if you ring the bell early, you can only have one marshmallow." The experimenter exited the room, leaving the plate with one marshmallow on the table in front of the child. The experimenter then recorded when the child rang the bell or ate the marshmallow. If the child held out for fifteen minutes, the experimenter returned and gave the child the doubled reward.

Few of the children were able to wait fifteen minutes. On average, they lasted just over six minutes. But there was substantial variation among the children. More than a quarter caved in less than two minutes; another 25 percent lasted more than ten minutes. Now, you might think that this variation doesn't signal any deep difference in the children's impatience. The study didn't control for the children's blood sugar. Maybe kids who'd skipped breakfast that morning were more likely to grab the marshmallow at hand. But Mischel and his coauthors found that this crude experiment on delayed gratification in four-year-olds predicted how the children would do many years later on their SATs. Four-year-olds who waited five minutes longer scored on average almost 300 points higher on their SATs in high school (126 points higher on the verbal portion, 171 higher on the quantitative portion). The marshmallow test also predicted how the kids' parents would respond to a survey conducted ten years after the initial experiment. The parents of kids who waited for the doubled reward were significantly more likely to say that their fourteen-year-old "exhibit[ed] self-control" and "pursu[ed] his or her goals." These parents were even more likely to describe their kids as being intelligent.

The authors were quick to emphasize that they don't understand exactly why there is this dispersion in patience.

> The causal links and mediating mechanisms underlying these long-term associations necessarily remain speculative, allowing many different interpretations. For example, an early family environment in which self-imposed delay is encouraged and modeled also may nurture other types of behavior that facilitate the acquisition of social and cognitive skills, study habits, or attitudes which may be associated with obtaining higher scores on the SAT and more positive ratings by parents. It also seems reasonable, however, that children will have a distinct advantage beginning early in life if they use effective self-regulatory strategies to reduce frustration in situations in which self-imposed delay is required to attain desired goals. By using these strategies to make self-control less frustrating, these children can more easily persist in their efforts, becoming increasingly competent as they develop.

The exact contributions of nature versus nurture are up for grabs. Moreover, impulse-control problems vary by context even within the same individual. I procrastinate horribly on my taxes, but I almost never put off grading final exams, even though I view both as approximately equally unpleasant recurring tasks. To be honest, there is still much to learn about for whom and when irrational impatience will rear its ugly head. Richard Thaler cautions that modesty is in order. "We are still closer," he said, "to the beginning than to the end in our understanding of what really lies behind hyperbolic discounting."

Most importantly, we, unlike pigeons, can organize our lives to over-come irrational temptation. As self-aware animals, we can take action to overcome it. What's distinctive about all these hyperbolic preferences is that they are temporary. Behavioral economists Ted O'Donoghue and Matt Rabin emphasize that our after-the-fact view of our actions differs from our initial impulses, noting that most humans "exhibit a tendency to pursue immediate gratification in a way that they themselves disap-prove of in the long run."

But even more important for our purposes is that people can regret their impending impatience *in advance*. With a sufficiently long hori-zon, people can realize that it's worth waiting a day to double their apple. Both before and after, we regret the short-horizon decisions. This is just what we saw in the pregnancy study, where both before and after preg-nancy mothers expressed different preferences for anesthesia than those they expressed when the question was a pressing one. It's our ability to regret in advance that creates the battle between our present more-patient selves and our future give-it-to-me-now selves.

Richard Thaler, of course, understands the implication of these pref-erence reversals. It is as if we have multiple selves—an intertemporal Sybil-like split personality. In his excellent book *Nudge* (coauthored with Harvard law professor Cass Sunstein), Thaler describes the titanic strug-gle between your inner Simpson and your inner Spock. Your Homer brain wants what it wants now. It hyperbolically discounts. In contrast, your inner Spock more analytically weighs the costs and benefits of delay and is systematically more patient.

But what's crucial to the new understanding is the timing of the con-flict. Preferences shift as time horizons shorten. Your inner Spock inhab-its your body first—it contemplates in December whether you should put off doing your taxes until the last minute. Our inner Homers come along later and systematically undermine the wiser choices, the best-laid plans and the resolutions we've made in the past.

Since our inner Homers can easily get the last word, it isn't really a fair fight. Not unless we put a thumb on the scale favoring our prior, more patient preferences—by committing our future selves to actions we make plans to enforce today. In *The Odyssey*, as his ship is approaching the irresistible but deadly Sirens, Odysseus commands his crew to tie his hands to the ship's mast, "You must bind me hard and fast, so that I cannot stir from the spot where you will stand me ... and if I beg you to release me, you must tighten and add to my bonds." Odysseus's prior patient self thus wins the fight with his future self by taking immediate action to incapacitate his future self.

In 1519 Hernán Cortés scuttled his ships upon arrival in Veracruz, Mexico, which required his vastly outnumbered six hundred soldiers either to defeat the Aztecs or to die trying, just as, nineteen centuries earlier, the Athenian general Xenophon fought with his back against an impassable ravine, having set himself and his men up so that they would have no option of retreat.

Hand-tying contracts have been with us forever. People enlist others' support to make sure they don't defect from their life plan in 12-step programs. Christmas club contracts and layaway plans have helped people save money for special purchases. Some people even use cursing jars, which enable them to stay true to a joint promise to stop cursing.

Behavioral economics didn't discover the possibility of commitment contracts. But the new learning of behavioral economics can help us to better answer three crucial questions about commitment contracts:

When do we need them?
When are we likely to use them?
How best can they be tailored to make them effective?

We've already made progress on the first of these questions. We need commitment contracts to help resolve the conflict between our inner Homer and our inner Spock. And we now know that these schizophrenic preferences are most likely to arise when we hyperbolically discount the future—that is, in contexts where we would give inconsistent answers to Thaler's apple questions.

But in order for our inner planners to benefit from commitment contracts, we need to acknowledge that we have a problem. We need to be sophisticated enough to acknowledge that our future selves are likely

to have different, temporary preferences. You might think that from repeated experience people would learn that they lack the willpower to follow through. After all, every year most of us make and then fail to live up
to our New Year's resolutions. But humans have an amazing ability to ignore the evidence from the past and convince themselves that this time
will be different. Thaler told me, "Just about everyone thinks they are
going to be a few pounds lighter six months from now, even though it
rarely happens."

Without sufficient sophistication, we won't take advantage of commitment contracts. (Wily Odysseus knew enough about himself to understand that he needed to incapacitate his future self.) So a first payoff
from this book is an increase in your own sophistication. Sure, we all
need commitment contracts to fortify the willpower of others that we de-
pend on. It's easier to see the willpower failings in others. And, to be
sure, your savvy can substitute for the lack of self-awareness of others.
That's why it would have been useful for me to require Simon Usborne
to make a binding commitment before agreeing to *re*-reschedule my
interview.

But it's harder to admit your own weakness. The hyperbolic tendency may cause us on one dimension or another to fleetingly give in to
temptation. Vertebrates may all have a hardwired tendency to hyperbolically discount the future. However, only humans are sufficiently self-
aware of their schizophrenic selves to parentally take action to restrain
their future freedom. The Robert Burns poem "To a Mouse, On Turning
Her Up in Her Nest with the Plough, November 1785," from which the
phrase "the best-laid plans of mice and men" derives, goes on in the final
verse to emphasize:

> Still you are blest, compared with me!
> The present only touches you:
> But oh! I backward cast my eye,
> On prospects dreary!
> And forward, though I cannot see,
> I guess and fear!

It is our capacity to look forward and backward that sets us apart.
Burns thinks that it must be blissful for mice to live constantly in the moment. But it is our ability as humans to look ahead and "guess and fear"

that is an informational prerequisite to the commitment enterprise. Armed with self-knowledge, our inner Spocks can take arms against our future preferences. Only humans know enough to seek out binding commitments.

Or not.

Ainslie went back into his pigeon lab and devised a diabolically ingenious experiment to test whether pigeons could take action to figuratively "tie their beaks." This time he started with ten White Carneau pigeons that again displayed a strikingly short-horizon impatience. When presented with the choice between immediate access to 2 seconds of food or access to 4 seconds of food if they waited just 3 seconds, more than 95 percent of the time these hungry birds couldn't keep themselves from pecking the smaller, sooner option. "The tendency to seek the shorter access to food when it was immediately available," Ainslie wrote, "remained strong after as many as 20,000 trials. However, casual observation through the oneway viewing lens often revealed the subjects to be pecking the chamber wall near the [immediate reward key], before pecking the key itself." Could this slight dithering peck of the wall reflect an inner conflict in pigeons as they tried to resist their inner Homer's call for immediate gratification? Yeah, right.

But Ainslie tested the idea by giving the pigeons a precommitment option. He added an extra, disabling key to the feeding apparatus. At the beginning of each experiment, a green key would light up for 7.5 seconds. If the pigeon left this green key unpecked, 12 seconds later the pigeon would be given the original smaller-and-sooner/larger-and-later choice. The only impact of pecking the green key would be to disable the subsequent immediate reward option.

So what happened? Over thousands of trials, 7 of the 10 birds showed no affinity for the disabling light. But three of the birds made systematic use of the disabling key, over time choosing to peck the lighted green key—ultimately up to 80 percent of the time. "What is remarkable," Ainslie explained, "is that some pigeons learned to peck a key at an earlier time, if and only if this made them unable to obtain the smaller reinforcement." Even though they had an overwhelming urge for impatience whenever presented with the possibility of an immediate reward, these three birds, when given an earlier "beak-tying" option, were smart enough to work against the expected future myopia.

To be honest, I still wasn't completely convinced by these initial results. Maybe these three pigeons just liked pecking keys. More specifically, I worried that they'd been so well trained to associate pecking with rewards that they had a hard time not pecking at any lighted key presented to them. But Ainslie foresaw this concern and ran another version of the experiment, one in which he flipped the default meaning of the disabling green key. Under this final version of the experiment, *not* pecking the initial green key disabled the temptation key. The green key became an enabling key that caused the subsequent red temptation key to light. If the three pigeons were simply peck-happy, we would expect them to keep on pecking the green key just as much as before. But these very same pigeons, when placed thousands of times in this new environment, switched course. Over time they pecked the enabling key less and less—ultimately less than 20 percent of the time, thus avoiding the temptation to be tempted.

At least some pigeons are sophisticated enough to use commitment mechanisms to avoid temptation. Of course, we are only talking about three birds. Still, the fact that 30 percent of Ainslie's birds were able to learn to use the commitment device is important. The fact that most birds could not bring themselves to use the commitment device is itself consistent with what we see in humans. Some people are smart enough to use commitment devices, but most of us do the same thing over and over, all the while expecting different results.

THE PROBLEM OF SOPHISTICATION

Matt Rabin is the cool uncle you always wished for but never had. Economics may be a dismal science, but Matt, who teaches at Berkeley, is downright playful. On a typical day, you'll see him tooling around campus in one of his many tie-dyed T-shirts. At the tender age of forty-five, he already has a few laugh lines around his eyes. His webpage has bizarre links telling you random facts (like the capital of Montana). You can also choose to see pictures of him "on a good hair day" or "on a bad hair day."

This playfulness is also characteristic of his writing. Most people who think about lack of willpower start by conjuring examples of procrastination—putting off till later unpleasant tasks that could be done more efficiently today. But Matt and Cornell economist Ted O'Donoghue have flipped things around to think about the opposite problem: of grabbing rewards too early when you'd be better off waiting. In fact, Matt and Ted had to go out and create a new word for the phenomenon, "preproperate"; it comes from the Latin *praeproperum*, which means "to do before the proper time." Traditional economic theory has no room for this kind of mistake. Rational economic actors weigh the costs and benefits of choices and tend to do the right thing.

We might not have had a word for it, but we've already encountered several examples of preproperation. Presented with the prospect of an immediate reward, the pigeons chose the early food, and humans switched to selecting the single apple. But Matt and Ted have done more than give a new name to an old willpower problem. By thinking carefully about preproperation, they have figured out that self-knowledge can sometimes be a bad thing.

Here's an example from Matt and Ted's 1999 article in the *American Economics Review:*

> Suppose you have a [free] coupon to see one movie over the next four Saturdays, and your allowance is such that you cannot afford to pay for a movie. The schedule at the local cinema is ... a mediocre movie this week, a good movie next week, a great movie in two weeks, and (best of all) a Johnny Depp movie in three weeks. Which movie do you [use your coupon to] see?

(By the way, Matt seems to have a thing for Johnny Depp. On Matt's webpage, if you click the link for the "good hair" day, you'll see someone who looks suspiciously like the matinee idol.) To make this stylized example more concrete, the article assumes that actually seeing one of these films would at the time of viewing increase your utility by 3 (for the mediocre film), 5 (for the good film), 8 (for the great film), or 13 (for the Depp film) units. Pretty obviously, we'd expect a rational consumer to wait for the highest-valued film.

But things change when we start allowing for hyperbolic discounting. Matt and Ted make an extreme but simplifying assumption that hyperbolic discounters cut in half the value of any future reward—whether it is 1 or 100 weeks in the future. So a hyperbolic discounter choosing between going in week 1 or week 2 would go now—because seeing a mediocre movie today is better than seeing a discounted good movie next week ($3 > 5 \div 2$).

Now, a naïve coupon holder will have the fortitude to forgo seeing the mediocre movie the first week, because he'll still prefer waiting for the Depp move three weeks later ($3 < 13 \div 2$). He'll also hold out the next week because even the lure of a "good" movie immediately will not be sufficient to outweigh the pleasure of getting to see a Depp film in two weeks ($5 < 13 \div 2$). But, sadly, the naïve discounter will not ever get to see Captain Jack Sparrow (or whomever Depp is playing this time around) swashbuckle across the screen. At week 3, our moviegoing hero will finally succumb to the same hyperbolic problems we've seen before. Suddenly, the immediate rewards of the third movie outshine those of even the great Johnny Depp at a week's remove ($8 > 13 \div 2$).

The naïve consumer doesn't realize how his tastes will change. In weeks 1 and 2, he is confident that he will hold out for the best of all possible movies—without realizing how the opportunity to see the third movie will look once he has the chance to see it immediately. If only the consumer were more sophisticated . . . matters would be even worse.

You see, a more self-aware consumer would realize how her tastes are likely to change over time. Even in week 1 she would realize that when the third week rolls around she will cave in and go to the third movie. Indeed, if she is sufficiently sophisticated, she'll realize that she won't even make it to week 3. She'll know that at week 2, she'd prefer to see the good movie today (that is, in week 2) rather than the great movie that she'd end up watching in week 3 ($5 > 8 \div 2$). And knowing this, she won't even make it to the second week. Here's a case where the good is the enemy of the great.

So if she is truly aware of her future preferences, she'll end up going immediately to the mediocre movie. Even though in week 1 she strongly prefers to wait to see Johnny Depp, she knows herself well enough to realize that she will never get there.

This is the great unraveling. Knowing that you're going to preproperate a little, it will sometimes make sense to preproperate a lot. Sophistication can be a bad thing, because it can lead to a sort of "aw, screw it" attitude where you just throw in the towel immediately. The movie-coupon hypothetical may at first seem a bit contrived. But it immediately helped me understand why I sometimes do truly gross things like eat two pieces of chocolate cake for breakfast. Ideally, I'd save the cake to share with my beloved spouse after dinner. But when I'm working at home, I realize in the morning that I'm very likely to cave in and eat a piece as an afternoon snack. And once I know that, it isn't long before I find myself at eight A.M. with chocolate icing on my fingers.

After this kind of ungodly binge, I used to ask myself, "Have you no shame?" But now I ask myself, "Why do you have to be so damnably sophisticated?" If I weren't so accurate in predicting my future preferences, I'd have been able to hold out longer.

When I mentioned this "sophistication as problem" idea to a colleague, he suggested it might even explain some instances of premature ejaculation. "A guy might want," he said, "to climax at the same time as his partner. But if he knows he's not going to be able to hold out that long, the equilibrium may unwind with him sexually prepropering."

The Delphic oracle's "Know thyself" line might still be good advice, but Matt and Ted have shown that knowing thy*selves* can sometimes backfire. Knowing the preferences of your future inner Homers can make it easier for your current Homer to run roughshod over your immediate behavior.

What do we want? Cake. When do we want it? NOW!

Sophistication mitigates the tendency to procrastinate with regard to life's burdens but, perversely, sophistication can also exacerbate the preproperative tendency to do things too soon. This is a depressing thought. Self-knowledge isn't all it's cracked up to be.

But the unraveling effect of sophistication needs to occur only if you can't use your knowledge to take arms against this sea of preproperation. "If you'd set up a deal where your wife would slap you for eating the cake," Ted joked, "you might have saved some for dinner." Commitment can help solve the problems of both procrastination and preproperation. Of course, a face-slapping commitment seems a bit harsh. Many of us opt for the gentler palliative of separating ourselves from the temptation. (I'm writing this on the third floor of my house—far away from the kitchen.) The power of even these soft commitments can be substantial. In the marshmallow tests, simply covering the marshmallows helped the four-year-olds hold out nearly twice as long. This "out of sight, out of mind" strategy is just one of a bewildering array of potential commitments you might make to increase your resolve.

2

Incentives Versus Commitments

In the spring of 1996, a third-year law student, whom I'll call Elizabeth, came to me with a pretty big problem. She was supposed to graduate in a few weeks but had yet to write her "substantial," a paper that is an absolute requirement for graduation. I am reluctant to take on students under these circumstances. If last-semester students turn in bad papers, it is very hard for me to fail them and be the ultimate cause of their not graduating on time. But Elizabeth had an interesting paper idea and an infectious enthusiasm, so I signed on as her supervisor.

Elizabeth is a short (five-three on a good day), big-boned blonde who names all her pets after restaurants. She has a tanned, healthy look that reminds me of one of my favorite camp counselors. But unbeknownst to me at the time, she has had a persistent problem with procrastination.

As Elizabeth's graduation approached, I began to get nervous that I had not received a first draft of her substantial. Normally, for serious papers of this kind, I mark up an initial draft and ask students to submit a second, "new and improved" version. But with about a week to go before final grades were due, I still had not seen anything. With just a few days to spare, she finally submitted a draft that, to my mind, was skeletal at best. The paper had an interesting idea, and it was clear that Elizabeth had done a bunch of research. But the ideas just hadn't been thought out, and the writing was unpolished. Under normal circumstances, this is not the kind of paper I would be willing to sign off on. This was just the train wreck I had worried about. But did I really want to stand in the way of Elizabeth graduating on the following Friday?

Elizabeth was somewhat desperate. She promised up and down that if I would just give her a passing grade, she would revise the paper over the summer and turn in a much improved final version before school started in the fall. I didn't know what to do. In the end, I accepted her offer—but with a twist. I gave her a "provisional low pass." The low pass is the lowest passing grade at Yale Law School. It is not a mark of distinction. But it was enough for her to graduate. And the "provisional" meant that I had the option of raising her grade at any point in the future if she submitted an improved draft. The twist was that I asked her to stand behind her promise by agreeing to give $100 a month to a charity of her choice if she didn't get the revised paper in by September 1.

The summer passed and I got caught up in other projects and teaching a new passel of supersmart students. To be honest, I forgot about Elizabeth and her paper promise. But a few years ago, a young colleague of mine sent me a bizarre email. He was getting ready to attend his ten-year reunion at Yale, and he was worried that Elizabeth was going to skip the reunion because she was scared of running into me.

You see, Elizabeth had never turned in that revised term paper. After graduation, she'd moved to Colorado and studied for the California bar while preparing for a judicial clerkship. After she passed the bar that summer, she still very much intended to revise the paper. But things didn't go according to plan. "I procrastinated; I kept not doing it," she told me recently. "I moved the files on the paper from the house we were renting to . . . the house that we now own, which was, like, eight years ago. For about two years my intention was to revise the paper. . . . And I kept paying $100 to a charity every month as per our agreement."

That's right. Month after month, for ten years, she kept on contributing to a charity—first to CWEALF (the Connecticut Women's Education and Legal Fund) and later to a local humane society. Twelve thousand dollars in all. When I heard about Elizabeth's story, I was aghast. A decade later, she was no closer to completing her paper and felt that she was on the hook for a lifetime of monthly penalties. I worried that I had done wrong, maybe even acted unethically in imposing this financial condition. I immediately passed on the word that all was forgiven. I released her from any continuing obligation to make charitable gifts, and I encouraged her to come to her reunion.

But Elizabeth's story is important because it gives us another clue in trying to understand commitment contracts and when they are more (or less) likely to succeed. Arguably, my attempt to get Simon Usborne to call failed because he could tell himself that he'd never really promised to pay for failure. His ambiguous reply ("Your phone will ring!") allowed his future self to reframe and diminish the import of blowing me off for a third straight interview appointment. The same excuse doesn't work for Elizabeth. She knew she'd made a specific commitment to pay. And pay she did. In a sense, she abided by her contract promise: turn in a revision or pay $100 a month. Month after month she chose the second option. So much for my clever idea. But the failure of this $12,000 penalty should make us ask, What exactly is a commitment contract?

OF CARROTS AND WHIPS

The phrase "the carrot and the stick" has become unmoored from its original meaning, which referred to the use of a stick to suspend a carrot in front of a mule or donkey. Originally the whole phrase referred to a contingent reward (the carrot was, in fact, given to the beast upon returning home to its stable after the day's work). But today the carrot by itself represents the reward, while economists think of the stick as a contingent punishment that is inflicted if the goal is not met. It might be clearer if we started talking about carrots and whips to indicate the threatened rewards and punishments that pervade our lives.

If economics is the science of means, then carrots and sticks are the core laboratory instruments. These contingent rewards and punishments are the basic building blocks of economic incentives. Economists see incentives all around us—and it's a good thing, too. To be considerate is to consider the way your actions affect other people. The beautiful thing about incentives is that they can help us be more considerate.

For example, take dog poop...please. It's a pain to pick up after your dog. But not picking it up is seriously inconsiderate. Poop not only fouls your neighbors' shoes, it is a public health concern, because little kids end up eating it and get sick with worms and bacteria. Tika Bar-On decided to do something about it. Dr. Bar-On, the chief veterinarian for Petah-Tikva, a municipality near Tel Aviv, has spearheaded a high-tech program to give dog owners better incentives to do the right thing. Under a pilot program in the neighborhood of Neve Oz, dog owners who deposit their dogs' droppings in special park trash cans can win dog food, leashes, and even colorful poop-bag carriers.

How can Dr. Bar-On tell Fido's poo from Fifi's? DNA testing. First, Dr. Bar-On enlisted twelve-year-olds to canvass the neighborhood door-to-door to ask dog owners to donate saliva samples of their dogs' DNA. And she followed up, with a one-of-a-kind "DNA-donating festival featuring music, performing dogs and a booth for saliva collection." This cajoling (plus the prospect of rewards) has led the owners of more than two hundred of the neighborhood's four hundred registered dogs to submit samples to the dog DNA registry.

Then Dr. Bar-On started rewarding the considerate. "Every week, we take ten to twenty samples of poop from the garbage cans," she told me, "and send it to the laboratory." When the lab finds a match, the owner gets a mixture of both tangible and intangible positive reinforcement. "We call them and send them an email and we congratulate them," Dr. Bar-On said. "And we also have a website where [we] publish the names of the people that collected their dog poop."

The end result is that "this neighborhood is much . . . cleaner now than before," Dr. Bar-On told me. She is now working to pass a new law expanding the reach of the pilot program—but switching from a carrot to a stick. She hopes to make DNA registration mandatory and to fine dog owners when their dogs' wayward poop is found on the ground. But even these sticks come with some sweeteners. "Maybe we can also find paternity of the dogs and also find the owners of the dog when we find a stray dog in the street," Dr. Bar-On said. "Instead of putting the microchip in dogs, we can do the DNA exam."

Without incentives, many of us may be prone to ignoring the pains and pleasures of others. Incentive contracts loaded with carrots and sticks can help us internalize what would otherwise be the externality of other people's happiness and sorrow. For example, if you want to give drivers better incentives to choose when to drive in London, the classic economic answer is to levy a congestion charge equal to the social cost their driving imposes on others. The congestion charge internalizes what would otherwise be an externality—so we should expect people to drive only when their private benefit exceeds the social cost.

DOLLARS FOR SCHOLARS

Thus, the teacher may say, "Read and I will give you some nuts or figs. . . ." With this stimulation the child tries to read. He does not work hard for the sake of reading itself, since he does not understand its value. . . . As his intelligence improves . . . his teacher may say to him, "Learn this passage or this chapter, I will give you a dinar or two."

Now, all this is deplorable. However, it is unavoidable because of man's limited insight. . . . This is what the sages meant when they said, "A man ought always to labor in the Torah even if not for its own sake! For doing it not for its own sake, he may come to do it for its own sake."

MAIMONIDES, COMMENTARY ON THE MISHNAH

Roland Fryer knows that improved incentives are not just about feeling other people's pain. Incentives can also guide people to make better choices for themselves. Fryer is a wunderkind freakonomics-style economist from Harvard. But what's really getting people's attention is his "dollars for scholars" incentive projects. In New York City, Fryer was appointed the school system's CEO (chief equality officer), and with the full support of the New York City Department of Education's chancellor, Joel Klein, he has started experimenting with the idea of literally paying students to get better grades.

One of Fryer's programs—called Million for the city's approximately one million students—rewards students with cell-phone minutes. In the pilot, which launched in 2008, twenty-five hundred students in seven middle schools were given Samsung flip phones already filled with 130 prepaid minutes. And they could earn extra minutes by showing up and behaving well in class, as well as doing their homework and getting good grades.

In 2008, Fryer's Rewarding Achievement program (REACH, for short), paid kids from thirty-one low-income schools substantial cash bonuses if they passed an Advanced Placement test: $500 for a score of 3, $750 for a score of 4, and an even $1,000 for a top score of 5. And REACH gives even stronger incentives to the schools themselves, doubling the cash payments for every additional student who earns a passing grade. The results for the first year were somewhat mixed: 19 percent more students received the top score of 5, and 13 percent more students took the test. But the number of students with passing grades (of 3 or higher) was essentially unchanged.

Roland, however, is undeterred. In Washington, D.C., his Capital Gains program is paying sixth through eighth graders from fifteen schools up to $1,500 a year for better grades, behavior, and attendance. Students can earn points worth $2 each for simply showing up to class. They also get a point for every class "in which they're not disruptive, profane, or disrespectful." And as with New York's REACH program, Fryer is directly incentivizing achievement with $20 for an A on any test.

Fryer remembers that when he was growing up, he would see counterproductive billboards saying, "Crack is whack, school is cool." But now he's trying to make achievement cool, by giving kids who perform well the chance to use a high-status item, a cell phone with MP3, video, and texting capabilities. What's more, the cell phones give the school a powerful way to stay in contact with the kids. Now teachers can text their students with assignments and reminders of upcoming tests.

The numbers aren't in on all these multimillion-dollar projects, but paying students for grades has already produced tantalizing benefits in Israel. In 2001, the Ministry of Education ran a large-scale randomized experiment at forty nonvocational high schools to see whether it was worth it to pay students $2,400 for a passing score on the Bagrut—an exam given at the end of high school. A passing grade is a prerequisite for university admission. Students who were offered the incentive were 5 percent more likely to earn a Bagrut than those who were not offered the incentive—although, surprisingly, the impact was somewhat inexplicably concentrated among girls offered the bonus.

Fryer's not sure whether his educational carrots will work in New York. But he is committed to following the data wherever it leads him. He told Steven Colbert, "If it doesn't work, we'll try the next innovation next year." But his working hypothesis is that carrots improve performance. In fact, at the beginning of the interview, Fryer put a twenty-dollar bill down on the table as a reward if Colbert asked a good question.

A SUBSIDY THAT PAYS FOR ITSELF

Of course, poorly structured incentives can backfire. In India, a government bounty on rat pelts apparently had the perverse effect of inducing

private entrepreneurs to breed rats. This is the law of unintended consequences applied to incentives. But not all unanticipated consequences
are bad. When Peter Cramton and I analyzed the impact of affirmative
action on the FCC's auction of thirty narrowband licenses, we found, surprisingly, that it increased the government's revenue by nearly $45 mil-
lion (more than 12 percent). The FCC granted minority-owned firms a
massive bidding credit, entitling them to pay only fifty cents on the dollar if they were the high bidder in the auction. You don't need a PhD in
economics to see that the bidding credit was a huge carrot to encourage
minority participation.

But how in the world could letting some bidders pay less increase
the revenue? The answer is that affirmative action changed the incentives of other bidders in the auction. Suddenly, the telecom companies
faced new competition. Even though the government lost money on the
licenses it sold to minority bidders, it more than made up for this with
the higher prices it received on licenses it sold to traditional telecom op-
erators. This effect was not subtle. In one example, an unsubsidized bidder (PageMart) had succeeded in outbidding all of its traditional rivals
by offering $76 million for a package of regional licenses. But then a
minority-controlled bidder, PCS Development, entered the fray, upping
the ante more than a dozen times before dropping out. PageMart still
ended up winning the licenses. But because of its altered incentives, it
paid the government $17 million more. Improving the incentive of tradi-
tional telecoms to pay a fair price wound up more than paying for the
government's cost of providing an affirmative action subsidy.

AN OFFER TOO BAD TO ACCEPT

Commitment contracts are all about incentives, but they operate with a
special kind of incentive. Ordinary incentives seek to guide prospective
choices. Commitment devices instead try to take prospective choices off the
table. (Following the Nobel Prize–winning economist Thomas Schelling, I
prefer the term "commitment" to "precommitment." All commitments
are made in advance, so adding the prefix "pre" is slightly redundant.)
This choice-destroying quality is easiest to see in figurative hand-tying
arrangements that disable disfavored future actions. Dieters engage in a
variety of disabling commitments. Some literally lock the refrigerator.

Others—more than 100,000 Americans each year—have their stomachs stapled, drastically disabling their capacity to consume substantial quantities of food at any sitting. After the procedure, the average stomach is reduced from about a quart to less than an ounce—and even after stretching out again, it may be able to hold only about a cup. Meta-analysis over fifty years shows that most patients lose more than 65 percent of their excess body weight after the gastric bypass procedure. But a minority of patients find ways to feed their caloric demons by eating more frequently—and by switching to liquids. Milkshakes and Mountain Dews can defeat even the smallest stomach.

Other disabling commitments have shorter-term horizons. Antabuse is the trade name for a drug (disulfiram) that helps people commit, for about a week, not to drink alcohol. Like the gastric bypass, the drug does not literally disable you from drinking. But Antabuse makes drinking alcohol immediately unpleasant. Within five minutes of ingesting alcohol, an Antabuse user will experience severe nausea and vomiting, along with all the other symptoms of a horrible hangover. Clinics have tried to use Antabuse to help alcoholics kick the habit. Just coming in once a week to a clinic to drink the Antabuse is effective at ensuring that the alcoholic will stay on the wagon. It is probably not a coincidence that Antabuse was originally manufactured by Odyssey Pharmaceuticals (named after the granddaddy of hand tyers). The problem is that (unlike Odysseus) people who want to drink simply stop showing up at the clinic.

A popular weight-loss version of Antabuse is alli (the over-the-counter brand-name for the drug orlistat), which partially blocks the body's ability to absorb fat. As with Antabuse, alli users who eat too much fat are immediately visited with unpleasant digestive side effects. In a clinical study, more than 20 percent of first-time alli users experienced "oily spotting, flatus with discharge, and fecal urgency." Wags call this an "alli oops." These problems sound a lot like the marketing nightmare Procter & Gamble endured with the fat substitute olestra (which the FDA, using a memorable phrase, originally said could cause "anal leakage"). But unlike olestra's side effects, the alli oops may be part of a learning process. The same study found that with extended use the forgoing problems went way, way down—from 20 percent in the first year to just 2 or 3 percent in the second year. Why? Because users come to learn that eating too much of fatty foods will bring dire consequences, so they stop doing it. At the end of the day, alli is probably not a silver bullet. Even alli's own press materials modestly claim that "if you could lose 10 pounds through dieting alone, you could lose 15 with alli." But a 50 percent improvement by physically disabling your body from digesting fat is a step in the right direction.

Disabling commitments aren't just for the other guy. I've used them to help me stop abusing the Internet. The first thing I do when I buy a new computer is to delete Minesweeper, having lost countless hours to the cursedly addictive game. I've even had a law school administrator set up a software program called Cybersitter on my computers to stop me from surfing the wrong parts of the Internet. The software was originally developed to help parents prevent their kids from seeing sexually inappropriate materials. I use it to stop myself from spending too much time reading about the NBA and other sports. I've given a law school administrator "parental" control over what I can see. And every few months, I ask the administrator to adjust the filter—adding sports sites that I'm wasting time on or opening up sites I need for legitimate research. For a while, the software was deleting every mention of sports-related words. I got used to seeing "tranation" instead of "transportation." It's embarrassing for a grown man to admit that he needs this help, but that hand tying has helped me write this book.

The point of all these disabling commitments is to take future choices off the table. But you can also take choices off the table by connecting them to sufficiently large financial incentives. Run-of-the-mill incentives merely try to "price" the impact of your behavior on others so that you can make the right decision. Any Econ 101 class will tell you that the optimal effluent tax on pollution will set the dollar amount equal to the social cost of the pollution. A factory confronting this tax will internalize the social cost and will pollute only up to the point where the private benefits are greater than the social costs.

Commitment contracts do not try to price the pains and pleasures of others. They try, instead, to create an offer that can't be rejected—or, for *Godfather* fans, an offer that's too good to refuse. Dr. Lisa Sanders, the Diagnosis columnist for *The New York Times Magazine,* recently celebrated the twentieth year of a smoking pledge she made with a friend who was also trying to quit. If either one smokes a cigarette, they promise to pay the other $5,000. They started by putting $1,000 at risk but increased their stakes as their wealth increased. Lisa Sanders didn't choose a penalty of $5,000 because she wanted her future self to seriously weigh the benefits and costs of smoking. The offer not to smoke is too good to refuse. Or, if you like, the offer to smoke is too bad to accept. Sanders wanted to create a substantial enough penalty that she would not have to think about it. Economic incentives are all about guiding people to make better choices, but commitment contracts are about removing and reducing choices. Giving your CEO stock options is an incentive contract; giving your friend five grand if you smoke a cigarette is a commitment. So far the contract has worked to perfection. Lisa and her friend have both been smoke-free for more than seventy-three hundred days. Even though they've long since kicked the habit, they see no reason why they shouldn't keep the contract in place—just in case they're ever tempted to backslide.

In Israel, so-called kosher phones are combining both the disabling and the financial forms of commitment. A subsidiary of Motorola has sold more than 100,000 specially modified handsets that disable users from accessing the Internet, sending text messages, or even taking photos; the stripped-down unit lets you only send and receive calls. A similar phone offered by Sprint Nextel in New York provoked litigation in 2006 when users, contrary to the marketing materials, found a way to use their phone for texting. But in Israel the disabling tool goes even further. The service provider, MIRS Communication, working with a rabbinic council, has blocked more than ten thousand telephone numbers for phone-sex and escort services. At Yale, I have my cybernanny block sport sites. But in Israel, you can have your rabbi help keep you on the straight and narrow path.

The phones' pricing plan also helps take choice off the table. As with some other calling plans, there is an incentive to call other members of the network. Calls outside the network are 9.5 cents a minute, while calls to other kosher phones are just 2 cents. But even more interesting is that any calls placed on the Sabbath cost a whopping $2.44 per minute, which outstrips even the most rapacious cell-phone plans. The system could have been designed to disable all calls on the Sabbath. But call disabling would block even emergency calls (which are allowed to save a human life). The purpose behind the penalty, like Dr. Sanders's $5,000 cigarette, is to keep people from even thinking about the possibility of making a nonemergency call.

My favorite verb construction in the Russian language is the difficult-to-pronounce *ne vz*, which means "don't even begin to," as in "don't even begin to think about parking your car here." That's kind of the whole goal behind commitment penalties. They are figuratively trying to stop you from even thinking about the possibility.

CARROTS AS COMMITMENTS

When my kids, Anna and Henry, were eight and ten, they were desperate for a TV. Our family was one of the few in their circle of friends that didn't have one. We had an eight-inch screen that could play videos and DVDs, but my poor son and daughter were without access to *The Suite Life of Zach & Cody, Blue's Clues,* and the like. They felt socially ostracized because they had difficulty answering crucial questions of the day, such as "Who lives in a pineapple under the sea?" I had been holding out because I'm somewhat addicted to bad TV and have wasted countless hours watching crappy shows. So being a TV-less house was a disabling commitment to keep me out of harm's way. But pressure was mounting from my beloved spouse, as well as from the kids, to break down and get a TV.

And I did, but with a twist (with my spouse's grudging approval). I offered to get my kids a TV and a PlayStation if they would first, without complaint, take my corporate finance class that fall. We had been studying corporate finance over the summer (yes, I am that weird), and I wanted to expose them to it at a higher level. I also wanted to see what I could get them to do before their probable descent into teenage rebellion.

It's natural to think of penalties as taking options off the table. But carrots can do the trick as well. The TV-and-PlayStation offer was too good to refuse. No kid in his or her right mind would pass up this deal. Or at least that was true for my kids. When they were in the middle of the course, it never really occurred to them to balk and stop coming to class, because they wanted so badly the tube at the end of the tunnel. The prospect of losing a truly spectacular carrot can also keep you from considering alternatives. Bribing your kids with electronic baubles, however, can quickly become a pretty expensive way of increasing commitment; you have to actually pony up and pay for commitment rewards that are too good to refuse.

But the CEO of Zappos, Tony Hsieh, has enhanced employee commitment by offering a contingent-reward offer that costs very little because most employees turn it down. Zappos, which was established in 1999, is now the nation's largest online seller of shoes, with gross sales of more than $800 million in 2007. Its meteoric rise in sales is all about its superb customer service. My co-blogger at the *New York Times* Freakonomics site, Steven Dubner, tells of the story of his wife's purchase of a pair of ill-fitting sandals. Zappos not only offered to refund the price and to pay for return shipping, it also offered to send her a $25 coupon toward her next purchase. Then, Dubner reported, in a modernization of *Miracle on 34th Street,*

> Zappos also offered to try to locate a pair of the sandals in her size from another vendor. (Hah! Sure, they will!) Fifteen minutes later, the company called my wife and told her they'd found her sandals, in her size, at another online merchant—"and," the Zappos clerk told her, "they're even cheaper at this other site!"

How can you fill a company with employees like this, who are willing to go the extra mile? One way is to bribe them to leave.

At the end of the first week of a four-week training course, new Zappos employees receive "The Offer": In addition to paying them for the time they've already worked, Zappos will pay them $2,000, no strings attached, to quit. We're used to giving employees bonuses to encourage them to stay on the job. But Zappos dangles a substantial "quit now" carrot that it hopes trainees will refuse.

In part, the company uses the offer as a screening device. If you're the kind of employee who prefers a quick two grand to the opportunity to work at a great company, Zappos doesn't want you. In part, Zappos uses it to credibly signal that working long-term for Zappos really is worthwhile. (I don't recommend that McDonald's offer $2,000 to its new trainees.)

But what's really interesting to me is the psychological impact of "The Offer." By turning down the bribe, trainees signal to themselves early in their Zappos careers that they value their jobs. The act of turning down the offer increases the employees' internal commitment to the company and to succeeding on the job. We've seen before carrots that were too good to refuse and sticks that were too bad to accept, but the power behind the Zappos offer is that it is a very substantial offer that people *can* refuse. By rejecting the offer, trainees learn that they value their future job even more. Rejecting the offer makes them feel better about their employer. And they are psychologically motivated to make good on that initial choice by working hard to succeed.

Traditional economists like to say that sunk costs don't matter. But the idea behind the Zappos offer is that a sunk opportunity cost (turning down the one-time bribe) can influence your future behavior. To reduce cognitive dissonance, people like their current choices to conform with their previous values. The forgone offer gains its power by creating a moment where new employees have to fess up to really valuing this opportunity. The TV show *Temptation Island* gleefully celebrated the failings of couples who give in to temptation. But the Zappos offer suggests that resisting temptation can create a virtuous cycle where you want to be the guy who continues to act consistently with that initial choice.

Of course, all this only works if most employees turn down the offer. It would be a pretty lousy idea if trainees left in droves after the first week. Zappos was worried about this too, so the "quit now" bribe was initially set at just $100. When almost none of the employees accepted, Zappos bumped the offer to $500 and then to $1,000 before raising it to its current level. But the amazing news is that only 2 or 3 percent of the employees cash out and accept the $2,000 offer. Traditional carrots can be expensive if they are successful. But Zappos has created a powerful carrot that is low-cost because it is so often unused.

On the face, the Zappos offer is a kind of anti-incentive. It gives people an extra reason to do something the company doesn't want them to do. But its success makes me wonder whether there could be an analogous anti-incentive on the stick side. Instead of bribing employees to quit, Zappos might have demanded that trainees pay $2,000 for the privilege of continuing their training. As with the bribe, this seems like a really stupid idea that will cause mass defections. But those who pay the training fee would have signaled that they value the job more than $2,000 and are likely to work hard to prove to themselves that they were right to pay. I like the Zappos version better chiefly because the company doesn't earn a profit from those employees who signal greater commitment. Health clubs and even Weight Watchers centers sometimes require up-front membership fees, in part to raise their users' sense of commitment: "I paid $500, I better use it." But it's a little unseemly that they're making money from the deal.

Stepping back, we can see in the table in figure 4 a deep symmetry between carrots and sticks. Just as there are carrots that are too good to refuse, there are sticks that are too bad to accept. These are the commitment devices that disable choice, because no rational future self would choose the other side of the carrot or the stick. Then there are traditional incentives that guide choice by either subsidizing or taxing a particular action. And most perversely, we can now add the anti-incentives, which generate choices that can help you learn how much you really care about something.

Figure 4. A Typology of Incentives and Commitments

A TYPOLOGY OF INCENTIVES AND COMMITMENTS			
	COMMITMENTS (DISABLE CHOICE)	INCENTIVES (GUIDE CHOICE)	ANTI-INCENTIVES (SIGNAL CHOICE)
CARROT	Offer kids TV	Dog poop prize	Zappos offer
STICK	$5,000 cigarette	Congestion tax	Pay to train

The perverse benefits of anti-incentives might also explain what happened a few years ago on I-15 near San Diego. In 1996, the main lanes of the I-15 were terribly clogged during the morning and evening commutes, while the parallel express lanes, reserved for cars with two or more passengers, were underutilized. San Diego responded to the stark and frustrating misallocation of resources by launching FasTrak, a pilot program that opened express lanes to solo drivers if they paid for access. Taking advantage of new transponder technology, the FasTrak system automatically charges solo drivers a fee that varies with the level of traffic in the express lanes. As described by University of Chicago professor Lior Strahilevitz:

When a driver approaches an entrance to the Express Lanes, he sees an illuminated sign specifying a certain toll for using the Express Lanes. The more traffic is in the Express Lanes, the higher the toll will be. Not surprisingly, the Express Lanes tend to be most crowded when the general lanes are most crowded, so there is a strong correlation between the toll charged and the congestion in the main lanes on I-15. When the solo driver uses the Express Lanes he must pass through a certain lane, a transponder mounted inside the car signals a ground-based sensor, and a deduction in the amount of the posted toll is automatically made from the FasTrak user's prepaid account. (Carpools have their own specially-marked lane, and no deduction is made from a FasTrak user who drives through this lane.) In 1998, the average tolls ranged from $1.95 to $2.26 per trip.

More than ten thousand drivers signed up for the transponders and started shifting traffic from the main lanes to the express lanes. During the rush-hour commute, traffic in the express lanes increased by about 15 percent, while traffic in the main lanes went down by about 3 percent. Solo drivers using the express lanes reported saving ten to forty minutes for every ten miles they drove on I-15.

But what's really interesting about FasTrak is its impact on car pools. Simple incentive economics would predict that carpooling would decrease. Some carpoolers would choose to drive solo once the new option was available. Others would be dissuaded from carpooling as FasTrak made the express lanes more congested. Yet, surprisingly car-pool use increased substantially after FasTrak was launched—by more than 20 percent during the peak commuting periods. Why would a more congested lane suddenly become more attractive? The idea of anti-incentives might hold the key. The FasTrak users, by showing their willingness to pay more than two bucks to ride in the express lanes, credibly signal to potential carpoolers that the express lanes have value. With the Zappos offer, I argued that an employee's own choice signaled value and thereby changed future behavior on the job. With FasTrak, the choices of others might signal value and thereby increase carpooling, notwithstanding the anti-incentive of increased congestion.

This typology of incentive devices is a toolbox that might be applied to modify any decision. For example, imagine that you wanted a friend to quit smoking. We might use any one of the techniques in figure 5 to help her quit.

Figure 5. Incentives and Commitments for Quitting Cigarettes

	APPLICATION TO SMOKING		
	COMMITMENTS	INCENTIVES	ANTI-INCENTIVES
	(DISABLE CHOICE)	(GUIDE CHOICE)	(SIGNAL CHOICE)
CARROT	$100,000 reward	$5 per cigarette not smoked	receive $500 to quit support group
STICK	$5,000 cigarette	$5 per cigarette smoked	pay $500 to join support group

A traditional economic incentive would be a reward (say, $5 per cigarette) for smoking less or a tax (say, $5 per cigarette) for smoking more. The more perverse anti-incentives—such as a pledge to pay the smoker $500 if she quits a cessation support group or requiring a $500 payment to join one—would ostensibly give her a disincentive to stop smoking. But by overcoming and acting against that disincentive, a smoker might enhance his or her internal commitment to the cessation process. Both incentives and anti-incentives are about guiding or channeling internal choice. The Zappos offer doesn't take the future option to quit off the table, it just makes it psychologically less palatable. When you wouldn't quit for $2,000, it seems foolish to later quit for nothing.

Alternatively, you could adopt one of the harder-core commitment strategies to take the choice to smoke off the table. You could try to physically eliminate the presence of cigarettes in your friend's life or hope that science develops a tobacco analogue to Antabuse or alli. Or, like Dr. Sanders, your friend could promise to pay $5,000 if she ever lights up another butt. Alternatively, choices can be taken off the table if somebody offers your friend an outlandish reward—say, $100,000 if she promises not to smoke. We'll learn that carrot commitments like this can produce dramatic success; but, you might reasonably ask, who's going to be there to put up such an outlandish reward?

THE CASE OF THE RICH UNCLE

On March 20, 1869, at the golden wedding anniversary of his parents, "in the presence of the family and invited guests," William E. Story Sr.

promised his fifteen-year-old nephew (William E. Story II) that if the nephew "would refrain from drinking, using tobacco, swearing and playing cards or billiards for money until he became 21 years of age," the uncle would pay the nephew the sum of $5,000. It's a cool coincidence that Story in 1869 and Sanders in 2009 both chose five grand. But $5,000 in Story's day is closer to $100,000 in today's dollars. The outlandish size of this carrot was more than your ordinary incentive. It was the kind of offer that no person in their right mind could refuse. And, as succinctly described in that famous chestnut of contract law *Hamer v. Sidway*, the nephew "assented thereto, and fully performed the conditions inducing the promise"—that is, the nephew accepted the offer and then refrained from engaging in the uncle's laundry list of wayward behavior.

When the nephew turned twenty-one and politely wrote to the uncle asking for his reward, the uncle responded:

Dear Nephew:

Your letter of the 31st ult. came to hand all right, saying that you had lived up to the promise made to me several years ago. I have no doubt but you have, for which you shall have five thousand dollars, as I promised you. I had the money in the bank the day you was twenty-one years old that I intended for you, and you shall have the money certain. Now, Willie, I do not intend to interfere with this money in any way till I think you are capable of taking care of it, and the sooner that time comes the better it will please me. I would hate very much to have you start out in some adventure that you thought all right and lose this money in one year. The first five thousand dollars that I got together cost me a heap of hard work. You would hardly believe me when I tell you that to obtain this I shoved a jack-plane many a day, butchered three or four years, then came to this city, and after three months' perseverance, I obtained a situation in a grocery store. I opened this store early, closed late, slept in the fourth story of a building in a room 30 by 40 feet, and not a human being in the building but myself. All this I done to live as cheap as I could to save something. I don't want you to take up with this kind of fare. I was here in the cholera season of '49 and '52, and the deaths averaged 80 to 125 daily, and plenty of small-pox, I wanted to go home, but Mr. Fisk, the gentleman I was working for, told me, if I left them, after it got healthy he probably would not want me. I stayed. All the money I have saved I know just how I got it. It did not come to me in any mysterious way, and the reason I speak of this is that money got in this way stops longer with a fellow that gets it with hard knocks than it does when he finds it. Willie, you are twenty-one, and you have many a thing to learn yet. This money you have earned much easier than I did, besides acquiring good habits at the same time, and you are quite welcome to the money. Hope you will make good use of it. I was ten long years getting this together after I was your age. Now hoping this will be satisfactory, I stop.

Truly yours,

W. E. Story

P.S. You can consider this money on interest.

Across the decades, this amazing letter speaks to us about the dual nature of commitment agreements. The carrot of $5,000 was more than sufficient to commit the nephew to clean living. But the prospect of paying the unwholesome sum of $5,000 felt very different to the uncle writing the letter in 1875 than it did to the uncle who'd magnanimously offered the reward in 1869. The uncle never did make good on his promise (or the interest) and died in 1887, when Willie was thirty-three.

Every generation of first-year law students learns about this case because the uncle's estate refused to pay off on Willie's claim, arguing that the uncle's promise was gratuitous. The estate argued that there was no legal consideration for the uncle's promise—no quid pro quo—because Willie had not given back anything of value to the uncle. In fact, Willie had personally benefited from performing his promise to "refrain from drinking, using tobacco, swearing and playing cards or billiards for money." If the estate's argument had won the day, carrot commitment contracts wouldn't be enforceable. But the court brushed aside as a red herring the fact that Willie might have benefited from sticking to his own commitment, and found that it was sufficient consideration that he had promised not to do something that he had a legal right to do in the absence of a contract.

The power of the law to implement commitment devices goes far beyond the enforcement of rich-uncle contracts. A crucial difference between a mere price incentive and a commitment penalty is deterrence. Another way to think about commitment devices is that they take choice off the table by deterring you from making the wrong choice in the future. And many, many legal rules are about deterring individuals from making the wrong choice in the future.

Guido Calabresi and Doug Melamed revolutionized legal thought in 1972 when they noticed that the law protects legal rights in two strikingly different ways—with what they called "liability rules" and "property rules." If we enter a contract for you to drill a well at my home, I have a legal right to your performance. But my right is merely protected by expectation damages that compensate me if you don't show up; at work here is a liability rule. To be sure, the prospective damages give you an incentive to do the job. But their purpose isn't to deter all breach. If it turns out that you have unexpectedly high costs of performance, you might be better off compensating me instead of trying to blast through the unforeseen bedrock. Contract damages don't try to take the breach option off the table, they just try to guide promisors to breach only when breach is efficient. On the other hand, my right to be free from your trespassing on my land is protected by property rules. The law doesn't ask whether your trespass was efficient. (If it's really efficient for you to gain access, you are free to offer to rent it from me.) It instead threatens to send you to jail if you willfully enter without my permission. Criminal laws in general can be thought of as social-commitment devices that take choices off the table. But it's not just jail cells that can deter. Sufficiently high fines and penalties—especially those that carry with them social disapproval—can deter people from breaking norms.

The problem is that penalties that are intended to deter can become mere prices. I am embarrassed to say that I have at times treated the fines for overstaying at New Haven parking meters as a mere cost of parking, and not as the violation of a "thou shalt not" law. Some small bars in the Netherlands have similarly responded to a 2008 smoking ban by charging patrons a "smoker's entrance fee." The fee compensates the pub owner in advance for the fine, but, as reported in the *Economist*, the smoking continues unabated.

Frank Easterbrook, the polymath University of Chicago law professor who is now a federal judge, has gone so far as to suggest that managers of multinational corporations have an ethical obligation to violate criminal laws if doing so would increase the firm's profits:

> Managers do not have an ethical duty to obey economic regulatory laws just because the laws exist. They must determine the importance of these laws. The penalties Congress names for disobedience are a measure of how much it wants firms to sacrifice in order to adhere to the rules; the idea of optimal sanctions is based on the supposition that managers not only may but also should violate the rules when it is profitable to do so.

But this tendency to convert commitments into mere prices can undermine the value of commitment. You've probably heard this somewhat sexist joke:

> Man: Would you sleep with me for a million dollars?
> Woman: Hell, yes—a million dollars is a lot of money.
> Man: How about for a hundred dollars?
> Woman: Do I look like a whore to you?
> Man: That's already been established, ma'am. We're just quibbling over price.

When you start pricing what was originally inalienable, you are implicitly putting a choice back on the table. In fact, the undermining effect of turning a commitment into a mere price might have been a root cause of Elizabeth's inability to hand in her paper at the end of that summer many years ago. When I spoke to Elizabeth recently, she recalled a distinct feeling of relief when I offered her the option of giving $100 a month to charity if she was late. In fact, by adding on the $100-per-month option, I might, perversely, have decreased the odds that Elizabeth would finish the paper on time. What I viewed as a commitment contract, Elizabeth almost immediately viewed as a mere price incentive.

Elizabeth understood that I probably intended the charitable contribution as an intensifier of her promise to perform, not merely a price of nonperformance. She confided in me that her feelings of relief and guilt were somewhat contradictory: "At the time, I did not choose to say to you, 'Okay, good. So as long as I contribute $1,200 a year to a charity in perpetuity, then we are cool. Right?' I didn't say that because I expected that you would say, 'Well, no. That's not what I meant.' [Leaving the availability of exit somewhat ambiguous] was enough for me. . . . There is a certain amount of guilt but not so much that it resulted in my redoing the paper." The human psyche has an amazing ability to simultaneously hold contradictory frames in its mind. Especially when the temporizing future self comes on the scene, it's possible to start viewing penalties as prices—and not even very large prices, at that.

Traditional incentives have the key advantage of flexibility in the face of changed circumstances. The upside of guiding future choice is that your future self still has the opportunity to change course. During the summer after Elizabeth's graduation, a loved one might have fallen gravely ill and needed Elizabeth's help. Under such circumstances, paying $100 a month for several months would clearly have been the right thing to do.

But the key disadvantage to traditional incentives (or anti-incentives) is that they may not be sufficient to overcome the present-biased impatience caused by hyperbolic discounting. Anti-incentives, like the Zappos quit-now bribe, are going to be effective only if you show yourself that you can initially resist. Offering Elizabeth a bribe *not* to write the first section of her paper by the Fourth of July would likely not have been effective, because Elizabeth was already looking for excuses to delay.

The problem with traditional incentives is that they give you flexibility even when there aren't changed circumstances. Each month, month after month, the temptation of putting writing the paper off just a bit more was too hard for her to resist. In the end, she paid more than $12,000 (and might still be paying to this day if I hadn't relieved her of her obligation).

But imagine what might have happened if she had made a larger lump-sum commitment—to pay, say, $10,000 if the paper wasn't turned in by Labor Day. A recurrent result in the hyperbolic-discounting literature is that transforming a small, repetitive tax into a large, one-time lump sum can be an incredibly effective way of overcoming present-biased preferences. For example, the same Ted O'Donoghue and Matt Rabin who taught us in the last chapter about the mysteries of preproperation have proposed using a lump-sum-commitment idea to replace the death-by-a-thousand-cuts cigarette tax. Listen in on how economists speak to one another in this quote from our nation's leading economics journal:

[S]uppose that instead of charging a $2-per-pack tax on cigarettes, we charged $5,000 for a picture ID that allows that person to purchase up to 2,500 packs tax free and made it illegal to purchase cigarettes without this ID. Such a policy change could help *all* types of consumers. All the 18-year-olds who are rationally deciding to become lifetime nicotine addicts would purchase the license. The 18-year-olds who instead end up paying $5,000 in taxes for a lifetime habit they did not identify as optimal when they started would not buy the licenses. (If there were concerns that this scheme would prevent optimal experimentation, we could also issue a one-time "learner's permit" allowing a person to purchase up to 10 packs of cigarettes.)

The value of these big, lumpier commitments is that they break hyperbolic discounters out of the cycle of confronting a series of smaller, irresistible temptations. Einstein defined madness as continuing to do the same thing over and over while each time hoping for a different outcome. Yet that, in a sense, describes what Elizabeth did when she (at least initially) kept paying the monthly $100 but continued in her conviction that the next month would be different. It describes my own behavior when I keep eating that extra piece of chocolate cake while nonetheless believing that in the future I'll be better.

So a first piece of advice: when traditional carrots and sticks that merely encourage you to stop eating that extra piece of cake are insufficient, you should consider disabling your future choice literally or with supercharged commitments that even the most impatient future self will not be able to ignore.

This is not a bad place to start. But it's really just the beginning. To be most effective, we need to learn not just when to use commitments but what kind of commitments we should be using. Commitments come in a bewildering range of shapes and sizes. In the next three chapters, we'll focus on three related commitment questions: To what? To whom? and With what consequence? "To what?" focuses on the exact form of the substantive commitment. If you want to lose weight, are you better off committing to keeping your weight down? Or is it better to commit to exercising more or eating less? And when, if ever, is it better to simply commit to get on the scale regularly and report your weight?

The "To whom?" question asks who else should be party to the deal.

Should you trust someone else to referee your commitment? Are you better off if your referee is making a reciprocal commitment? And who should benefit if you fail?

Finally, "With what consequences?" asks what the consequences of failing to keep your commitment should be. We already know that commitments can be structured as carrots (that are too good to refuse) or sticks (that are too bad to accept). But is it sometimes better to use carrots *and* sticks? And should you rely on assets or honor as motivation? Ideally, all three of these sets of questions will be answered simultaneously to produce the optimally tailored personal commitment. Sadly, the state-of-the-art behavioral knowledge is nowhere close to pulling off that kind of everything-at-once tailoring—especially when the perfect-fit commitment will vary depending on the context and the person. But I hope to convince you that dozens of behavioral experiments can serve as guides to point us in the right direction.

3

Losses Loom Large

The first serious test of commitment contracts occurred in England more than thirty years ago. A psychologist named Robert W. Jeffrey led a research team that contacted overweight middle-aged men and asked them if they were interested in participating in a fifteen-week program to try to take off 30 pounds. To qualify for the study, the men (ages thirty-five to fifty-seven) had to be free of diabetes and heart disease and had to report drinking fewer than six alcoholic drinks a day. Most important, they had to be willing to lose some of their own money if they failed. At the beginning of the study, participants put at risk cold, hard cash; they earned it back week by week only if they stayed on track losing weight.

You already know that I'm a fan of the big stick. Over these past few years I've put more than $50,000 at risk. As of now, I still have yet to forfeit a single penny. But proud as I am of my own success, this anecdote doesn't say anything about how large a stick is sufficient to change behavior. Jeffrey wanted to see what works better, putting at risk $30 or $300. So he created an experiment that offered randomly selected participants either the chance to enter into a $30 commitment contract (which would pay participants back $2 at the weekly weigh-in if they lost 2 pounds) or a $300 commitment contract (which would pay them back $20 each week if they lost 2 pounds). For many people, this would be the difference between a mere incentive and a more substantial commitment. In today's dollars, this is like testing the difference between putting about $130 and $1,300 at risk.

Jeffrey predicted that "larger contracts would tend to discourage program participation but that among those participating in the study, larger contracts would produce larger weight losses." This simple prediction captures one of the great and pervasive tensions in crafting commitment contracts: the commitments that would end up being most effective are often not the ones we are willing to commit to in the first place. It is sometimes difficult to get your present self to agree to participate in a commitment contract where the stakes are very high. In the economics literature, gearheads would say you have to meet the "participation constraint." If a wise sovereign could simply impose welfare-enhancing commitment devices on the public, she could probably greatly improve the quality of life for many individuals. But physicians have to obtain "informed consent" from subjects to ensure that they are willing to shell out money if they lose.

As Jeffrey feared, the take-up rate for the $300 contract was lower than for the $30 contract. But the difference was not as large as you might predict. When presented with the opportunity to put at risk $30, 57 percent of subjects signed on the dotted line; of those subjects asked to cough up $300, close to half (47 percent) agreed to participate.

And of those who signed up, guess who worked harder to lose weight? On average, the men who had $30 at risk lost 23.4 pounds, while the average weight loss for the $300 group was 32.1 pounds—more than the program's 30-pound goal!

Of course, we should be a little worried that it wasn't the commitment contract that really drove this result. It might be that only the smaller group of men who were really sure they were going to lose 30 pounds were willing to put $300 at risk. But the beautiful thing about a randomized trial is that it is possible to ask what the impact was on the average person who just arbitrarily received the $30 or the $300 offer. If we assume that the people who turned down the offers didn't lose any weight, it is easy to estimate that the average weight loss within the group that received the $300 offer—including those who chose not to participate—was greater than for the group that received the $30 offer (15.0 versus 13.4 pounds). So on average, you'd be better off receiving (and accepting) the $300 offer. Being given the option of putting $300 at risk was better not just because of the larger average weight loss, but because that loss occurred at a lower cost.

In fact, the success rate of the $300 contracts was so high that less money was forfeited per pound lost by those who entered into those more expensive contracts. Of the men who put $300 at risk, 71 percent succeeded in losing 30 pounds. In the empirical literature on weight loss, this is an extraordinary success rate. In contrast, the success rate for the $30 contracts was still high, but only 38 percent. I crudely calculated that the men offered the $300 sticks lost an average of more than 5 pounds for every dollar they forfeited, while the $30 men lost only 3.5 pounds for every dollar forfeited. And because men were randomly assigned to these two groups, we can be confident that both groups of offerees were equally susceptible to success or failure.

But all is not peaches and cream. A year after the study ended, Jeffrey went back and found that both groups had regained most of their weight. The $300 club was still ahead, but not by much (a 13.8- versus 11.8-pound net loss). To me this result is not so troubling. Who said commitment contracts ever have to end? Lisa Sanders happily celebrated the twentieth anniversary of her substantial commitment not to smoke. And I still happily put $500 at risk every week. The fact that the commitment devices lose effect when they end is a weakness only if you think that the purpose of the device is to educate you. But Odysseus didn't tie his hands to teach himself how to avoid temptation. If he had gone back to the Sirens a second time, unencumbered by restraints, we would expect him to have jumped overboard.

By the way, I haven't mentioned it yet, but while he was at it, Jeffrey also tested the impact of having a third, randomly selected group of overweight men put $150 at risk. As you'd expect, he found both intermediate participation rates (50 percent) and success rates (47 percent). But what's really interesting is that putting a moderate amount at risk produced better results after one year. The men who had entered into either the $300 contract or the $30 contract were more likely to have regained most of their weight a year later. In contrast, the men who entered into the $150 contract ended up, on net, losing 2.5 more pounds a year later. The more extreme contract seems to have created more of a psychological backlash. When the men finished the more extreme contract, they were more likely to revel in their new freedom by returning to their high-caloric ways.

This possibility that smaller sticks can produce better long-term results is reminiscent of a classic series of "forbidden toy" studies. In these experiments, researchers asked preschool children not to play with an attractive toy (for example, a battery-powered fire engine) under threat of either mild or severe researcher disapproval. The severe threat delivered by the experimenter was the statement "If you play with [that toy], I would be very angry. I would have to take all of my toys and go home and never come back again." In contrast, the mild threat merely warned, "If you play with it, I would be annoyed." In these studies, both disincentives were equally effective at inducing short-term compliance. But interestingly, the researchers found that weeks later—after the children had been expressly told that the prohibition had been lifted—the children who had been threatened with the more stringent stick were more likely to play with the once-forbidden toy than the children who had received the less stringent threat (77 percent versus 33 percent). As in Jeffrey's studies, the more severe stick seems to have produced a behavioral backlash in the postincentive period.

These studies suggest that we may want different kinds of sanctions for short-term incentives. When the hope is to instill a behavioral change that will outlive the incentive, it may be advisable to choose a smaller stick. The trick is to select a contingent punishment that is still large enough to change short-term behavior, without triggering the longer-term backlash. If the stakes are too low, we may not succeed in fulfilling the initial commitment. If the stakes are too high, we might not be willing to enter the contract or we might be less successful in the long run. There is only so much we can learn from these fairly small experiments, but one takeaway lesson is that in setting the size of the stick, we need to think about the possibility of longer-term, even lifetime, commitments and attend to the Goldilocks problem of making the stakes neither too small nor too big.

THE FRAME'S THE THING

Which is better, Spanish or Bulgarian wine? Your answer could turn on all kinds of things—for example, the particular vineyard or whether you are offered a Bordeaux or a cabernet. But which wine you prefer in the future is also likely to turn on which you initially have. Two Dutch economists at random gave college students at Leiden University either a bottle of Spanish Torre del Arco or a bottle of Bulgarian cabernet sauvignon and then asked them whether they would like to switch. Since most of the students who initially received Bulgarian wine were chosen by chance, on average they should have had the same preference for the Bulgarian wine as the students who initially received the Spanish wine. But people were very reluctant to give up what they already had. Students who were given the Bulgarian wine were twice as likely to say that they preferred Bulgarian wine as those who were given the Spanish wine. And students who were at random given the Spanish wine were twice as likely to say that they preferred Spanish wine.

People hate giving up something they already own. So do monkeys. A couple of years ago, my friend Keith Chen confronted six tufted capuchin monkeys with two economically equivalent gambles. The monkeys were placed alone in cages where they could obtain food by going to one side of the cage or the other. On one side, a researcher displayed two slices of an apple; on the opposite side, a researcher displayed one apple slice. If the monkey approached the two-slice side, 50 percent of the time the researcher would give the monkey the two slices. But randomly, 50 percent of the time, the researcher would remove one of the slices and give the monkey just one slice. Alternatively, if the monkey approached the one-slice side, 50 percent of the time the researcher would give the monkey the one slice. But randomly, 50 percent of the time the researcher would bestow two slices.

Hyperrational monkeys should be indifferent between these two alternatives—each gives a fifty-fifty chance of one slice or two slices. Less rational monkeys might just simply gravitate toward the side displaying the larger initial amount of food. But surprisingly, Keith found that every one of the monkeys, after they figured out the nature of the game, systematically preferred going to the side where they could potentially win an extra slice rather than going to the side where they could potentially lose. Overall, the monkeys went to the one-slice side 71 percent of the time. Like the college students who didn't like giving up their wine, the monkeys didn't want to risk giving up their apple slices.

Behavioral economists call this loss aversion. And this aversion to losing things provides a great reason why Jeffrey was smart to focus on the incentive effects of stick contracts. If he had had a large enough grant, he could have tested whether a $30 or a $300 carrot was sufficient to induce weight loss. To a traditional economist, a $30 stick should have exactly the same incentive effects as a $30 carrot. In both cases, you are $30 richer if you succeed than if you fail. But simple experiments on people's willingness to trade suggest that losses loom larger than gains when it comes to incentive effects. Indeed, they loom twice as large.

When Richard Thaler was teaching at Cornell, he and the Nobel Prize–winning economist Daniel Kahneman gave students a Cornell mug (with a retail price of $5) and asked them at what price they would sell the mug. The students' average answer was that they would sell it only for more than $7. But what's really interesting is that when the experimenters gave another randomly selected group of students money and asked them at what price they would be willing to buy the mug, the average answer was less than $3.50.

Given these results, it is a short step to thinking that people will work twice as hard to avoid the pain of a stick than to secure the pleasure of an equally sized carrot. If you want to get a die-hard member of the Red Sox Nation to quit smoking, it will be more effective to threaten to take away his season tickets (if he has them) than to dangle the prospect of season tickets in front of him (if he doesn't).

Moreover, sticks are the gifts that keep on giving. It's a pretty obvious point, but punishments have to be imposed only when people fail, while rewards have to be granted when people succeed. This basic asymmetry has a huge impact on the relative cost of carrots and sticks as incentives. If the incentives work, carrots cost a lot more than sticks. It can cost Roland Fryer a lot of money to pay schoolchildren to behave in class if they end up behaving. So carrots as incentives can be really expensive, but carrots as commitments—the kind of hypercharged positive incentive that no reasonable person could turn down—will normally be exorbitant. Unless you have a truly rich uncle, it's normally just not fiscally feasible to go beyond carrots as incentives. In contrast, the *threat* of a punishment that works doesn't actually cost anything. The $500 that I put at risk last week is still in my pocket and can be used again this week to keep my weight below 190 pounds.

One of the amazing things about stick contracts is that putting more at risk can lower your expected forfeiture. For example, imagine a smoker who would have a 10 percent chance of quitting with $50 but an 80 percent chance of success if he put $200 at risk. Raising the stakes would reduce his expected forfeiture cost from $45 to $40. As long as the probability of forfeiture falls faster than the increase in the amount of risk, you can expect to lose less money when you put more at stake. In other words, as long as the probability of success is what economists call "elastic," you can save money by risking more money. As the stakes get sufficiently high, the probability of success becomes inelastic and the expected cost of commitment will rise. I put $500 at risk every week to lose weight—but I shouldn't expect to reduce the probability of forfeiture much if I upped it to $5,000. So at some point, truly high stakes will lead to higher expected losses. The bottom line is that commitment sticks that take choice off the table can impose fewer costs than mere incentives, because in equilibrium they induce a lower rate of failure. This is just what we saw in Jeffrey's original weight-loss experiment, where the $30 incentive ended up costing participants more than the $300 commitment.

BETTING ON CARROTS

Kevin Volpp remembers hearing his six-year-old twin daughters arguing over who would get to wear a particular dress. "Thea said, 'If you let me wear the dress today, I will let you wear it the next two times,'" Kevin recalls. "And Anna said no. And then Thea said, 'Okay. How about ten times?'" Kevin is a PhD economist as well as a physician at the University of Pennsylvania who still maintains an active practice seeing patients as a general internist.

His daughters' impatience reminds me of the pigeons and hyperbolic discounting. But the story reminds Dr. Volpp of the difficulty of getting people to take care of themselves. "It was immediately clear," he told me, "that the discounted value of using that dress umpteen times in the future was just trivial compared to the value of being able to do this right now. I don't want to equate adults who have a very serious health problem with kids arguing over clothes, but I think it is true that the salience of immediate frequent rewards is just really, really powerful." So when Dr. Volpp saw his daughters negotiating over the dress, he worried about how he could negotiate with patients to get them to take better care of themselves. The problem is that they have to incur a current cost to get future rewards. A lot's at stake. Dr. Volpp knows that 40 percent of premature deaths could be prevented if we could change people's present behavior, particularly with regard to how much they eat.

So as an economist and physician, Dr. Volpp wanted to devise a treatment that would give patients immediate incentives to lose weight.

He could have followed Jeffrey and tried to convince people to shell out a lot of money in advance and then pay them back at weekly weigh-ins. But he worried that he wouldn't be able to convince people to participate. People can't put up money they don't have. Volpp told me that Jeffrey's deposit contracts are "just not applicable for low-income populations where a lot of the problems with obesity lie." For Volpp, the participation effect loomed large.

Instead, Volpp crafted a carrot incentive for weight loss that in many ways represents the cutting edge of what behavioralists know about how to tailor incentives. He recruited healthy but overweight men and women from the Philadelphia VA hospital and asked them if they would participate in a randomized experiment. Half of them were assigned at random to a weight-monitoring program. At the beginning of the study, these veterans spoke with a nutritionist one-on-one for an hour about diet and exercise strategies that would help them lose weight. They were also given a free scale so that they could record their weight. And they agreed to weigh-ins at the end of each month. To make sure that they showed up each month, Volpp offered them $20 to weigh in, regardless of their how much they'd lost.

Volpp didn't expect this informational approach to be very effective. "We wanted to put the bulk of our resources," he said, "into the incentive intervention and not into imparting information, because our sense is that most people, by the time they are middle-aged, have a pretty good sense of what they are supposed to eat and what not to eat. And we see it as more of a problem of motivation."

All the action was in the incentive group, the other half of the randomly assigned participants. They received all of what the first half did (the one-on-one session, the free scale, and the $20 for each monthly weigh-in) *plus* an added financial incentive to actually lose weight. The vets in this group were given a chart showing on a daily basis the weight they needed to reach if they wanted to lose sixteen pounds in sixteen weeks. They were told "to weigh themselves each morning before eating or drinking and after urinating, record their weights, and call in their weight to the project staff by noon." But the key difference was monetary. "They basically were told," Volpp explained, "each day for the sixteen weeks that you . . . are at or below your [weight-loss] goal, you will be eligible for an incentive."

The incentive was the chance to take part in a daily double lottery that gave participants a 1-in-100 chance of winning $100 and a 1-in-5 chance of winning $10. The vets in the incentive group who came in at a weight that was at or below that day's goal got the chance to play their unique two-digit number in the lottery. Here's how it worked: At the beginning of the study, the participants in the incentive group chose a two-digit number—say, 27. Then each day, Volpp's team held a lottery randomly generating a two-digit number. As described in the team's 2008 *JAMA* article:

> If the first digit generated was a "2" or the last digit was a "7" (which has approximately a 1 in 5 chance) and the participant met his/her daily weight loss goal, he/she would win $10. If the randomly drawn number was "27" (a 1 in 100 chance), the person would win $100.

On average, the lottery provided a $3 daily incentive to lose weight, so that participants who succeeded fully day after day could expect to make $336 over the course of the sixteen weeks—or even more if they got lucky. But why the fancy lottery in the first place? Why not just pay them $3 a day if they made their daily goal weight?

The answer has to do with the innovative ways Volpp was trying to balance competing behavioral effects. From his daughters, he knew that he wanted to maximize the immediate effect of the carrot. "Behavior-change efforts are often futile because those changes might help you sometime in the future, but the average person has trouble making that future reward relevant today," Volpp said. "Cash reward programs offer a chance to change behavior and get a reward right now."

But he also had good reason to think that lotteries provide a bigger motivational bang for the buck than nonlottery carrots. The reason is that people tend to give more weight to small-probability events—especially those that for some reason become salient. Even though a 1 percent chance of winning $100 should have an expected value of one buck, subjects in study after study work harder to get a lottery ticket with an expected value of $1 than they will to get a dollar. This overweighting of small-probability events can also motivate people to work harder to avoid statistically unlikely punishments. My kids tell me that they'd work much harder to avoid a 10 percent chance of having their heads shaved than to avoid a 100 percent chance of having to cut their hair two inches. Our family doesn't do either—in part because of the participation constraint, but especially because it seems arbitrary and capricious to subject your loved ones to such severe probabilistic punishments.

"I think that probabilistic rewards probably work better because they have the built-in entertainment value," Volpp said, "but they only work if there [are] frequent opportunities for interaction." The problem is that a pure 1 percent lottery doesn't give the subjects enough of a taste of success to satisfy Volpp's immediacy concern. Volpp and his coauthors came up with the hybrid lottery to thread the needle—giving the subjects a frequent enough taste of winning to whet their appetites for the $100 "grand prize."

"What we were really trying to do," Volpp said, "was to give daily feedback." Volpp's team would send each participant text messages reporting how much money he had earned that day. For those who failed to meet their goals, Volpp even tried to cultivate an immediate sense of regret: the text messages would also tell them how much they could have made. Even though the incentive was the carrot, by framing the absence of the carrot as a loss, they tried to deploy some of the benefits of a stick contract as well.

Like Jeffrey's studies before it, Volpp's experiment was fairly small, with just nineteen vets in the control group and nineteen more assigned to the incentive group. But randomization even for this sample produced similar subject attributes in the two groups. The central difference was in the weight-loss treatment. After sixteen weeks, the control group had lost on average 3.9 pounds. In contrast, the incentive group had lost an average of 13.1 pounds. The probabilistic carrot more than trebled the amount of weight lost, and this difference was statistically significant. But these carrots were expensive. Ignoring the (substantial) costs of administering the program, the incentive group lost about .05 pounds (that's five one-hundredths of a pound) for every dollar received as a carrot. This is in sharp contrast to Jeffrey's study, which found that subjects lost on average 3.5 pounds for every dollar forfeited.

The Jeffrey and Volpp studies involved very different populations and time periods, so in some ways it is quixotic to make direct comparisons. But still, the stark difference in success rates and amounts of money transferred are at least suggestive of the power of loss aversion as a motivating tool. The threat of losing $300 was sufficient to induce 71 percent of Jeffrey's participants to lose thirty pounds in fifteen weeks.

The carrot of potentially earning $336 in the Volpp lotteries induced only about half (52.6 percent) of the participants to lose the goal weight of sixteen pounds in sixteen weeks.

CARROTS *AND* STICKS

But why does it have to be one or the other? Why can't we have our cake and eat it too by combining carrots and sticks to reward those who succeed and punish those who fail? The great advantage of the carrots-and-sticks approach is that it can harness the benefits of both rewards and punishments. Like Jeffrey's method, it can take advantage of loss aversion—the fact that we really hate to lose our own money—but can do this without running into the participation constraint. The lure of the third-party subsidy can bribe us to take on the risk of losing some of our own money.

Volpp saw the power of this approach and has been testing it, as well. In fact, the very same weight-loss study that tested the impact of the hybrid lottery included a third group of subjects who, at random, were confronted with a carrot-and-stick incentive. Vets who were assigned to a "deposit-contract" group were given the option to put their own money at risk *and* not only to receive their deposit back but, in addition, to earn extra dollars. "We told them," Volpp explained, "they could put down as little as a penny a day or up to $3 a day and then we would match that amount one for one." So at the beginning of a thirty-day month, vets in this group could deposit up to $90. And if they succeeded in meeting all of their daily goals and this weight loss was independently verified at the monthly clinic weigh-in, they could walk away with $180 for the weight loss (plus $20 more for coming to the weigh-in). To squeeze all of the potential motivational juice from the scheme, Volpp also diverted any of the forfeited deposit money from the vets who failed to take off weight to those who succeeded in their weight-loss plan. The vets knew that all forfeited money would be divided equally among deposit-contract participants who lost twenty pounds or more.

"What a deposit contract does is create a way where people put their own money at risk," Volpp said. "Once they put the money down, loss aversion is a powerful motivator. Nobody likes to lose money. And we turbocharge this where we match what their deposits are, to make it even more powerful, to make it even more likely people will try really hard to lose weight."

But (as they say on TV commercials) wait, there's more! Volpp also gave the deposit-contract group an additional flat reward regardless of how much of their own money they put at risk. "We also provided," Volpp said, "a direct payment of $3 if you met your weight-loss goal." Plus, there was an additional $50 for losing at least 20 pounds in the sixteen-week period. So all in all, a vet who succeeded in losing weight for a month was assured of receiving at least $90 and might receive as much as $320. Ninety percent of the vets in this group decided to put some of their money at risk. And the average daily amount they put on deposit was just above $1.50 a day—meaning that with matching and the $3 direct payment, the average vet would receive about $6 a day for meeting his or her weight-loss goal. The direct payment and the one-for-one matching made the subsidy for the deposit-contract group roughly twice the size of the expected subsidy for the lottery-incentive group.

This mixture of a mild stick and a certain, nonlottery carrot did even better than the lottery carrot. While the average loss for the lottery group was 13.1 pounds, the deposit-contract group lost almost a pound more (14.0 pounds) toward their goal of 16 pounds; again, this is much better than the control group, which with advice and a free scale averaged only a 3.9-pound loss. Moreover, there was less variation in the carrots-and-sticks group. Unlike the lottery and the control groups, none of the deposit-contract vets had gained weight at the end of sixteen weeks.

Again, there is only so much a study of fifty-seven vets can tell us. It can't speak directly to the dozens of other ways the study might have been structured. In fact, Volpp worries that they made the carrot component too lucrative. "I am not sure that we needed to do a direct payment," he confided. "We were worried that people wouldn't put deposits down and then we basically have a nonintervention, so I think we probably overcorrected in that regard." He is following up with a study that takes out the direct payment (but leaves the one-for-one matching) to see whether it can still produce superior results. I'd like to see a study that tried to maintain the benefits of having a lottery carrot. Instead of dollar-for-dollar matching, we might offer participants the following deal: if you put $3 at risk and lose the weight, we'll give you your money back, plus the chance to participate in the hybrid lottery ($100 grand prize, $3 expected value). The key question is whether the carrot of the lottery would be enough to get people to bear the threat of a $3 daily loss if they failed to lose the weight. If overweight people can bring themselves to participate, there is good reason to think that the combined immediacy of loss aversion and lottery attraction would be a powerful motivation to trim down.

ADDICTION FOR SALE

On February 22, 2008, James Hurman, a thirty-year-old up-and-coming advertising executive from Auckland, New Zealand, made commitment history when he posted a short video on YouTube offering to sell his smoking habit to the highest bidder:

> I've smoked cigarettes for twelve years and I've tried all the usual ways to quit smoking. Now that my wife Annabel and I are pregnant with our first child, it's time to give up once and for all. I've created a listing on the New Zealand online auction site trademe.co.nz, and on Monday, 31 March 2008, the highest bidder will receive a contract . . . in which I hand over my right to smoke to them, and agree to pay them a forfeit of NZ$1000.00 per cigarette that I smoke at any time following the auction's closure.

James had unsuccessfully tried to quit several times before. This was his way of getting serious. "I was sitting on this airplane," he told me, "and I was thinking about the guy that sold the million pixels and I was thinking about how he essentially managed to generate money for something that had no value whatsoever. And so, I was thinking about ways that I could make money selling something that has no value. And that is how I thought about selling my smoking habit. So originally it was actually a moneymaking ploy. But pretty soon, thinking about it further, it became not about the money at all and that I would give the money to charity and it became suddenly about smoking." He promised to donate the proceeds from the auction to the Cancer Society of New Zealand.

It's an arresting phrase, "selling my smoking habit." In one sense, James was selling to the highest bidder the right to be the recipient of any forfeiture payments. But in another sense, James was selling his right to smoke. Before signing the contract, he had an unfettered right to smoke whenever he wanted. And smoke he did. He estimates that he had smoked fifty thousand cigarettes before he entered this contract. (For the skeptical, his website, smokinghabitforsale.com, includes a video clip from a bar where he shows off his ability to throw a lit cigarette toward his mouth and catch it in midair. I'm convinced that he was a hard-core smoker.) But after the contract, he certainly no longer has an unfettered right to smoke. If infractions can be detected, $1,000 per cigarette for the rest of his life is a pretty substantial deterrent. "I didn't sneakily try to leave room for myself to get out of it, if you know what I mean," he said. "I made it really binding for myself. I think someone could wait until I smoked a pack and get $20,000 out of it."

James's attorney was, in fact, worried that the lifetime contract was too draconian. "He said it would be more prudent to make it for three years, or five years," James said. "But I really didn't want to do that because that kind of, for me, that sort of defeated the purpose. I wasn't doing it for publicity. I was doing it out of a genuine kind of desire to quit smoking forever." The contract did limit his liability to instances of "voluntary" smoking—because James was, frankly, worried that someone might coerce him to smoke against his will.

The auction and website generated a fair amount of press when James went public. Local news shows publicized his attempt and debated whether it would work. Thousands of people visited his website, and ten people bid on the auction. Several potential bidders questioned James about the details of the contract. For example, James explained that he had excluded cigar smoking from the contract: "There are ceremonial occasions, like the wetting of babies' heads, etc., at which it's customary. I don't really like cigars and have only ever had three, so I don't see them as a threat to my health, but I would like to be able to smoke them when it's customary to do so."

On other issues, he made clear that the contract was quite expansive. On the auction website, he told bidders, "There are no allowances for a drunken smoke." And: "In the spirit of the idea, one puff would need to count as 'smoking a cigarette' and thereby require payment of the full $1,000."

But there was a general concern that it might be difficult to catch James after the fact if he continued to smoke from time to time. James had made no special provisions for enforcement. He didn't post a bond with a bank. He also didn't give anyone else an incentive to narc on him. "It crossed my mind that people could buddy up," James said. "Like, say, if somebody who hadn't bought the contract was to see me smoking, then they might go to people who did buy the contract and say, 'Hey, I saw James smoking, so why don't we split it fifty-fifty.'" But in the end James, as an ad exec, relied on publicity to keep himself in line. As he put it, "There's no way of knowing for sure whether I ever smoke again. I guess that given the very public nature of this campaign, you can count on everybody watching me closely and dobbing me in if I falter."

A little more than a month later, on March 31, James smoked his last cigarette before signing his commitment contract with Kent Pearson. Kent had won the auction for a mere $300 (and James paid off on his first commitment by promptly donating the proceeds to the Cancer Society). "I thought the auction would be a little bit higher than what I got," James told me. "I thought it might have been more like $1,000. But I really had no idea. How do you put a value on something like that? So, at the end of the day the money that I got for it, the donation for the charity, was really a small part of it."

The auction revenue might have been low for two radically different reasons. Bidders could have been convinced that the commitment to pay $20,000 a pack ensured a very, very low chance that James would ever fail. Or bidders could have figured that they would never be able to collect from James even if he did continue to smoke. They might have reasoned that they wouldn't know if he smoked late at night at home, or that they wouldn't be able to prove it in court, or that they wouldn't be able to collect on a legal judgment. It's interesting that the high bidder, Kent Pearson, was a coworker of James's at Colenso BBDO. Kent would know if James really had been a smoker in the past. And he'd also have the inside track if James kept on smoking ten cigarettes a day.

Regardless of the amount of the winning bid, James still had a powerful incentive to quit. And quit he did. James had been under the contract for over a year when I spoke with him down under. His wife had their first baby, Tripp Sander Hurman, on July 21, 2008, and James was proud to report that he has been smoke-free since he signed the contract. "The interesting thing about this time," he said, "was that when I gave up, I didn't experience the cravings like I had done before when I tried to quit. For some reason, they kind of weren't there." Something about the contract made him lose the desire to smoke. So without nicotine patches or gum, James went completely cold turkey. I believe him, in part, because of the very public nature of his commitment and, in part, because he was happy to take a nicotine test to prove that he is clean.

James Hurman never went to college, but his idea responds powerfully to one of the central problems Dr. Volpp was worried about. Volpp knew that loss aversion tends to make commitment contracts with sticks more effective than incentive contracts with carrots. But Volpp was concerned that he couldn't get people to put enough money at risk. Volpp was worried about the participation constraint.

James's innovation might solve this conundrum. James chose to donate the proceeds of his auction to charity. But imagine the possibility of keeping the proceeds from the auction. A friend of mine from college, whom I'll call Jennifer, is deeply embarrassed about her nail biting, which she believes hurts her professionally. Jennifer is reluctant to write a commitment contract because she is afraid that she'll fail. But she might put $1,000 at risk as a commitment to stop gnawing on her nails, and then, like James, she could auction the rights to be the recipient of any money forfeited on the contract. The proceeds from the auction would be the carrot for her to enter into the commitment contract, and the threat of losing the $1,000 would be the stick to make sure she follows through.

The Hurman auction is ideally suited to respond to the problem of sophistication. Behavioral economists tend to think that to benefit from commitment contracts, you need to (1) have a problem with willpower, and (2) be sophisticated enough to know that you have a problem with willpower. People who naïvely think that they will have more fortitude in the future—to quit smoking, eat less, and so on—just won't believe that it's necessary to go through the hassle and risk of making a real commitment. But the Hurman auction compensates you in advance for taking on the hassle. The very hyperbolic discounting that makes people give in and smoke or eat to excess is likely to make the dangle of immediate cash especially tempting.

Moreover, people's naïveté about their future willpower will make the offered compensation even more attractive. Imagine again a naïve Jennifer who considers putting $1,000 at risk as a commitment to stop biting her nails for the next ten weeks. Because Jennifer naïvely believes that she will have much more willpower in the future, she may believe that she has less than a 10 percent chance of losing her stake. But a more objective bidder in the auction might believe that she would have a 20 percent chance of failure—and offer her $200 up front to take on this commitment. Jennifer's overoptimism and impatience will help overcome her resistance to commit. The amount bidders are willing to pay in these auctions can provide powerful information to the person making the commitment about their likely probability of success. If a bidder is willing to pay Jennifer $700 for the chance to gain a $1,000 forfeit, Jennifer should realize that the market thinks she has a pretty small chance of success.

Michael Abramowicz and I are working on an academic paper where we are spinning out more than a dozen variants of the Hurman auction. For example, in one version, an overweight Rush Limbaugh might auction the right to be paid $10,000 if he does not lose a certain amount of weight. But instead of having the auction bidders bid on how much they would be willing to pay for this right, Rush might demand $5,000 now and have the auction bidders bid on what his goal weight should be. It would be a pretty safe bet that Rush couldn't safely end up weighing 40 pounds, so an auction winner at this insubstantial weight would be virtually assured of being paid the forfeiture. But in a weight-bid auction, other bidders would bid up the goal weight until there was roughly a 50 percent chance of success (which is what bidders would demand if they were going to have to part with $5,000 on the prospect of earning $10,000). Rush could set the probability of success that he wants and then let the market decide what a reasonable goal is. Weight-bid auctions can respond to the problem that dieters routinely set unrealistic goals. We want the quick but unsustainable fix of taking off too many pounds, too fast, in ways that almost ensure that we'll pack them back on again.

THE FRAME'S THE THING (PART II)

In some forms of the Hurman auction, it will be difficult to tell what is a carrot and what is a stick. We talked about an overweight Rush being paid $5,000 today to take on the risk of forfeiting $10,000 if he fails to lose weight. But to make sure that Rush will pay if he fails, some bidders will want Rush to post the $10,000 in advance. Under this approach, Rush would need to post a $5,000 deposit in advance (his $10,000 stake minus the $5,000 compensation) and he would receive $10,000 as a return of his stake if he ended up meeting his goal. This is starting to sound a lot like a traditional wager. And it is. In fact, in England, the betting agency William Hill adopts just this approach when it takes weight-loss wagers. For example, Graham Trow won a bet that said he couldn't lose twenty-eight pounds in fifty-six days; he initially paid in $68, and he ended up earning $1,900 for his efforts.

In one case, Rush is being paid $5,000 up front (but has to pay $10,000 if he fails); in the other he is shelling out $5,000 (but gets paid $10,000 if he succeeds). The difference in the timing of the cash flows can limit Rush's ability to participate. He simply may not have $5,000 to put up right now. But the difference between these pay-me-now or pay-me-later contracts can have important differences in how we frame carrots and sticks.

In our home, we use a mixture of carrots and sticks in our child rearing. My beloved spouse prefers carrots, while I tend to opt for sticks. But we both see that one form of incentive can easily transmogrify into the other. Once a child is used to getting a carrot, its absence can feel like a stick. And the absence of a stick can feel like a carrot—it's so pleasant when the pain stops.

But behavioral psychologists know that we shouldn't be indifferent to framing. Imagine that it is October 23, 2002, and you are Vladimir Putin. You learn that, as *The New York Times* later reported, "40 Chechen guerrillas wearing masks and camouflage and firing automatic rifles stormed into a crowded theater in Moscow where a popular musical [*Nord-Ost*] was playing" and have taken 600 hostages. The guerrillas are heavily armed and are demanding the withdrawal of Russian forces from Chechnya.

What should you do?

An important experiment posed a question eerily similar to the *Nord-Ost* dilemma. In the experiment, all the subjects were asked to imagine that the United States is anticipating the outbreak of an unusual Asian flu epidemic that will likely kill 600 people. Two treatment programs are being considered, and the subjects were asked which they favored. Half the subjects were then given the "loss" frame. They were told:

If Alternative A is adopted, 400 people will die.

If Alternative B is adopted, there is a one-third chance that no one will die and a two-thirds chance that all 600 will die.

The other half of the subjects were given the "gain" frame. They were told:

> If Alternative A is adopted, 200 people will be saved.
>
> If Alternative B is adopted, there is a one-third chance that all 600 will be saved and a two-thirds chance that no one will be saved.

Even though the two choices are identical, subjects time and time again tend to respond to the framing with dramatically different results. When presented with the loss frame, 78 percent of the subjects choose the riskier strategy (two-thirds chance that all 600 will die); but when presented with gain frame, only 28 percent of subjects choose the riskier strategy (two-thirds chance that no one will be saved).

People are risk-averse when choosing between two different rewards but risk-preferring when choosing between two different sticks. One way to think about this bizarre result is that loss aversion is so strong that people are willing to roll the dice to give themselves some chance of avoiding a large loss.

This simple experiment (which has been replicated in different formats and across different types of subjects) suggests that Putin might have responded differently to the *Nord-Ost* siege depending on whether the decision was framed in terms of lives saved or lives lost. People are risk-averse with regard to gains. So if the siege decision was framed in terms of lives saved, it might have seemed more prudent to continue negotiations (even if some hostages would be killed each hour), rather than having police storm the theater (with the attendant risk that all the hostages would be killed).

OF LOTTERIES, INSURANCE, AND OVERWEIGHTING

The result of the Asian flu study—that people are risk-averse with regard to gains—seems to flatly contradict the earlier discussion that lottery carrots are more effective. But this tension can be resolved by seeing the framing effect as simply competing with a separate tendency of people to overweight small-probability events. Because of the framing effect, it is possible that a certain $100 carrot would provide a larger motivation than a 50 percent chance of winning $200. But the overweighting effect can reverse the result. A one-in-a-million chance of winning $100 million also has an unbiased expected value of $100, but if people *feel* as if they have a 1-in-a-10,000 chance of winning, they will treat the lottery as having an expected value of $10,000. Even with risk aversion, the lottery can provide a bigger motivation than a certain $100 carrot.

Judging solely from his actions, Putin is more likely to have framed the decision in terms of lives lost. After a two-and-a-half-day siege, Russian security forces filled the theater with a debilitating gas and then stormed the building. More than one hundred of the hostages (and fifty of the guerrillas) were killed, primarily from the effects of the knockout gas.

Framing choices as gains or losses is not just about policy decisions over whether to storm the theater or to distribute a flu vaccine. Framing can dramatically change the choices people make in their everyday lives. Peter Salovey has used framing to improve people's choices concerning both detection and prevention of disease.

I should disclose that Peter is now kind of my boss as provost of Yale University. But before he took on this administrative role, he worked as one of the country's great psychologists. He is one of the celebrated scholars who figured out how to measure and think about "emotional intelligence."

Because of Salovey, health-care organizations also have more effective messages to promote everything from HIV screening and mammograms to the use of infant car seats and even regular exercise. Salovey's central idea is that detection behaviors are psychologically riskier than prevention behaviors. When you have a breast exam, you might actually find out at the end of the procedure that you have a disease. In contrast, a prevention behavior—such as using sunblock—may be a bit unpleasant, but it exposes the user to very little psychological risk. Applying the results of the Asian flu experiment, Salovey developed a simple hypothesis:

The perceived risk (of finding an abnormality) could make loss-framed messages more persuasive in promoting detection behaviors. However, prevention behaviors may not be perceived as risky at all; they are performed to deter the onset or occurrence of a health problem. Choosing to perform prevention behaviors is a risk-averse option—it maintains good health . . . [B]ecause risk-averse options are preferred when people are considering benefits or gains, gain-framed messages might be more likely to facilitate performing prevention behaviors.

Peter thought that emphasizing the carrot of better health would induce more prevention behavior, while emphasizing the stick of poor health would induce more detection behavior. But Peter did more than theorize: he went out and tested his hypothesis in the real world—to see if, in fact, stick-framed messages worked better to promote detection and carrot-framed messages worked better to promote prevention.

He and his colleagues began by recruiting women who worked for a large telephone company. Female employees who needed but had not had mammograms were invited to view a fifteen-minute videotape on breast cancer and mammography. Half of the women were assigned randomly to view a video titled *The Benefits of Mammography;* the other half viewed a video titled *The Risks of Neglecting Mammography.*

Peter's hypothesis won hands down. The stick-framed message was much more effective at inducing women to act. After twelve months, 66.2 percent of the women who'd viewed the stick-framed video obtained a mammogram, compared to just 51.5 percent of the women who had viewed the carrot-framed video.

These results were reversed, however, when Peter went to the beach to investigate a literal ounce of prevention: applying sunscreen. Peter and his coauthors recruited 217 sunbathers on a public beach and randomly assigned them to read either carrot- or stick-framed brochures about sunscreen and the prevention of skin cancer. After reading the pamphlets, the sunbathers were given coupons for free sunscreen samples. But the neatest part of the experiment came next. About half an hour later, Peter sent a sunscreen vendor to the beach and watched which of the sunbathers actually turned in their coupons.

Salovey's prediction worked again. For this low-risk prevention behavior, the carrot frame was more effective at inducing participation: 71 percent of the sunbathers who'd read the carrot-framed pamphlet used the coupon, while only 53 percent of those who'd read the stick-framed pamphlet chose to lather up.

Salovey's experiment on relative risk suggests a rule of thumb for when it is best to frame an incentive as a carrot or a stick. Framing benefits as carrots is more likely to be effective when the desired behavior is viewed as reducing risk, but framing benefits as a reduction in sticks is more likely to be effective when the desired behavior is viewed as increasing risk.

We can also use incentive and commitment contracts to reinforce the preferred frame. For example, in Salovey's experiment, giving sunbathers the certain carrot of a coupon for free sunscreen might have reinforced the carrot frame of the pamphlet's message that using sunscreen increases gains. Analogously, a contingent punishment for failure to have a mammogram would reinforce the stick frame of the pamphlet's message, that taking the test reduces your losses. But using stick contracts to reinforce a loss frame will work only if we can overcome the ever-present participation problem.

A larger take-home point is that the tailoring of commitment contracts affects not only how effective the contract will be in changing your future behavior but also your willingness to enter into the contract now. To be successful, commitment contracts must appeal to your present self and simultaneously induce your future self to change its likely behavior. Time and time again, we will find that these two requirements are in tension.

ESCALATORS, DO-OVERS, AND THE DISADVANTAGE OF CASH

A related tension concerns how commitments should respond to the possibility of repeated failure. Lisa Sanders's $5,000 commitment creates a huge incentive for her not to smoke again. But the second cigarette costs her nothing. In contrast, James Hurman chose a constant $1,000 for each and every cigarette he smokes.

Behavioralists going all the way back to B. F. Skinner have experimented with schedules of punishments and rewards that vary over time as a function of success and failure. In particular, there is increasing evidence that it is often useful to reset the goals after an initial failure so that subjects have a chance to qualify again for a carrot (or avoid a stick).

Indeed, it shouldn't surprise you that Volpp and his colleagues incorporated a reset into their own weight-loss system. In his sixteen-week study, participants were given a "fresh start" if their weight at the end of the month ended up being greater than their goal weight for that date. In the weight-loss context, resetting is crucial to make sure that people don't have an incentive to starve themselves at the last minute to make an unreasonable goal. Resets also encourage continued participation. Regardless of a past failure, you still have some skin in the game to make progress.

Planning for failure with resets is particularly important when working with drug addicts, where inducing even short-term abstinence is hard-won. Stephen Higgins, a physician at the University of Vermont, has conducted more than a dozen experiments teasing out how different schedules impact addicts' behavior. In many of these studies, heroin addicts are offered escalating rewards for clean daily urine samples. The rewards start with vouchers worth $2.50 and can increase over a twelve-week period to more than $30, but they reset back to $2.50 if the addict produces a dirty sample (or if he is a no-show). In other studies, Higgins has shown enhanced abstinence using escalating lotteries, where addicts get additional daily draws from a fishbowl if they stay clean.

In 1996, he tested the impact of alternative incentives on smokers who were asked to abstain from smoking for just five days. Twenty randomly selected smokers were given about $10 for each day a breath test indicated abstinence (measured by a carbon monoxide level less than 11 parts per million). Another twenty subjects were given escalating carrots, which started at $3.50 and escalated by 50 cents a day, with a $10 bonus for three days of abstinence. The smokers in the fixed-$10 group were more likely to start off abstinent but were also more likely to fall off the wagon, while the smokers in the escalating group who started off clean were much more successful at sustaining their abstinence over the five days. Escalating incentives work, but only if the initial carrot is large enough to grab people's attention. In a world with limited budgets, there just may not be enough money to escalate from a large enough starting point.

A five-day study of forty smokers is not going to give us a definitive answer, but Higgins has had enough success in getting people to stop abusing cocaine, marijuana, alcohol, and amphetamines that it seems worthwhile to explore whether escalating carrots can help with weight loss and other potentially less addictive behaviors.

Of course, it's not feasible for the carrots to escalate forever. But the good news is that once someone turns over a new leaf, it is easier for them to maintain their good behavior with de-escalating carrots. For example, Kim Kirby, a psychologist who is now the director of the Treatment Research Institute, found good results in a twelve-week study of cocaine addiction. The program started off giving participants a larger carrot (a $30 voucher) for every clean test and later, after the ninth successful test, de-escalated by giving them the voucher only for every third clean test.

Parallel questions of escalation and de-escalation are at play with stick contracts as well. Penalties can easily be crafted to escalate after repeated failures. Small initial penalties that potentially reset after a period of renewed success can help induce more participation. At the very least, we should be wary of stick incentives that de-escalate too precipitously—like Lisa Sanders's smoking commitment, in which the penalty drops to zero after the first failure. Planning for the possibility of failure can increase the probability of success.

Finally, it is useful to consider whether the carrot or the stick should be denominated in the form of cash. The addiction studies paid their subjects with vouchers redeemable for goods and services for the simple reason that the researchers didn't want to help fund junkies' efforts to buy more drugs. The problem with vouchers—like gift cards—is that sometimes they are redeemable only for things you don't really want, and therefore can provide less of a motivator than cold, hard cash, which is the universal gift card, redeemable for anything that can be bought or sold. To neoclassical economists, tangible goods or services can never be better carrots than their cash equivalents. From this perspective, the almost $1 billion spent by firms on tangible sales incentives is inefficient relative to what might be accomplished with straight cash bonuses.

But behavioralists are finding that at least in some contexts, tangible goods and services can spur more motivation. Cash bonuses that are used to pay bills are quickly forgotten, but a tangible reward—especially if it's a luxury item—is there to remember every time it is used. An indulgence that you couldn't justify buying for yourself can be justified if it is won as part of an incentive program. "Consumers often feel the guiltiest about the things that provide them with the highest pleasure," says Ran Kivetz, a marketing professor at Columbia who has empirically tested the relative preference for cash equivalents. People who would never dream of spending money on a trip to Hawaii can nonetheless be extremely motivated to win this prize through hard work. Instead of feeling guilty about purchasing an indulgence, if you win it you can brag about it to your friends and coworkers.

For example, in one study Kivetz told a sample of travelers waiting at airports that as a token of the airline's appreciation they would be able to choose between either receiving $2 immediately or participating in a lottery where they would have a 1-in-a-100 chance of winning a dinner-for-two certificate with a maximum redeemable value of $200, which could "be used to dine at any of the top 30 leading gourmet restaurants in the U.S." He found that 84 percent of the travelers chose to gamble on winning the dinner certificate. But what's interesting is that Kivetz also asked another randomly selected group of travelers to make a parallel choice between receiving $2 immediately and participating in a lottery where they have a 1-in-a-100 chance of winning $200 cash. A neoclassical economist would have to expect that a higher proportion of travelers would prefer the cash lottery. After all, $200 in cash could be used to buy a fancy dinner or anything else costing $200—and there isn't the risk that all or part of the certificate would go unused. But Kivetz found that travelers were not as willing to participate in the cash lottery: only 65 percent opted for it (a statistically significant shortfall), indicating that at least in a lottery context, a luxury prize could be preferred to the universal solvent of money.

A parallel study at the University of Chicago showed that people were actually willing to work harder for massage certificates than for their cash equivalents. Scott Jeffrey, a management sciences professor who now teaches at the University of Waterloo, offered staff volunteers $10 to participate in a "Word Prospector" competition where they could win additional prizes for creating "as many four- to six-letter words as possible using the letters of a ten-letter target word." (For example, if the target word was GARGANTUAN, they could earn prizes for coming up with answers like "gnat," "grant," or "grunt.") The control group was offered cash incentives: participants who scored above the 20th percentile earned an additional $2, while those who scored above the 50th, 80th, or 95th percentiles earned bonuses of $10, $30, or $100, respectively. A second randomly selected group of staffers were offered noncash indulgences: scoring at the 20th percentile or better earned "a fancy candy bar worth $2," while higher-percentile scores earned coupons with parallel face values for goods like massages at a local spa. Again, a traditional economist would predict that the cash incentive would be more powerful. Surely, some of the competitors would not like massages or would not want to go through the hassle of setting up an appointment at a spa. Unsurprisingly, a substantial majority of the participants indicated in a survey that they would prefer the cash reward. But what is surprising is how the staff members actually performed. Competitors who were offered the massage incentives worked harder than those offered the cash incentives; they answered substantially more questions and earned scores that were 47 percent higher. The noncash incentive produced better results. We might worry that the staffers offered massage prizes just happened to be better at word-jumble games, but random assignment is a pretty powerful technique to make sure that the only systematic difference between the groups was the type of incentive offered.

Economists are trained to put a price on everything. Our profession's first instinct is to monetize. So it was natural for me to suggest that Simon Usborne should pay money if he didn't call me. To an economist, payment is pain. But Scott Jeffrey's studies suggest that financial penalties might not always be the best way to get people to take choice off the table. Ray Romano, in his documentary *95 Miles to Go,* makes "mind bets" with himself where he will commit to shipping his golf clubs home if he fails to make par. We're so used to exchanging money for goods and services that financial penalties might be easier for people to commodify. Yet the result that some tangible goods make more powerful carrots suggests that giving up tangible forms of consumption might be a more powerful stick. Bail-bond dealers understand this when they take as collateral clients' possessions that have very little market value. Used dentures or even a defendant's cat cannot be sold for much if the defendant skips town. But the bond dealers know that the potential loss of prized possessions can be a powerful motivator to induce someone to show up for their court date. I have a close friend who put at risk her $5,000 annual contribution to her alma mater as a way of backing up a weight-loss commitment. Not being able to support the school she loves is a much bigger deterrent than "merely" forfeiting $5,000 in cash.

It's hardly a surprise that carrots and sticks change behavior. But not all changes are worth the candle. In 2007, Eric Finkelstein, a health economist for RTI International, published a fairly straightforward test of carrots for losing weight and found only a slight impact. Over a three-month period, compared to a control group of unincentivized participants, overweight university employees who were offered $14 per percentage point of weight loss lost only 2.7 pounds more, and employees offered $7 per percentage point of weight loss lost only 1 pound more. The differences were statistically significant but hardly much to write home about. The chance to win up to $140 for losing 10 percent of your body weight just didn't resonate with people nearly as much as Volpp's lottery offer (which took off an average of more than 13 pounds in four months for an average cost of $262). The larger lesson here is that details matter. True believers may rush in with a firm conviction that any old carrot or stick can improve the world. But attending to specifics is shown in study after study to be a better recipe for success.

4

That Nagging Feeling

In the fall of 2001, as a new school year was starting, I had a hankering to connect more socially with Yale's supersmart law students. I also wanted to get back into running shape. I thought I'd try to kill two birds with one stone by offering to go running with students. In the past, I had allowed a student charity to sell the rights to "Run with Professor Ayres" at its annual charity auction. This pleasure of running with me didn't go for as much as "Drink Beer and Play Guitar with Professor Ayres" nights, but even without the lure of beer, Yale students were willing to pay for the right to run with me.

In that fall of 2001 I thought I'd flip things around. Instead of selling the right to run with me, I offered what I called a "charitable bribe." Any day of the week (except Sunday) on which at least one student showed up at my house at seven A.M. to run, I would contribute $10 to the charity of that student's (or those students') choice. I was giving them the chance to get some exercise, talk to me, and make some money for their favorite charity.

In one sense, the incentive worked pretty well. Over the course of the next couple of years, I got to know two students particularly well (and I've stayed in touch with them to this day). Moreover, I got into good shape. Students showed up only about a third of the time, but I didn't know exactly which days they would. Once I'd gotten up and put my kids in a jogging stroller, I'd go ahead and run even if no students had shown up that particular day. The running incentive worked a little bit like Bill Russell's shot blocking for the Celtics. Russell once admitted that he blocked only about 5 percent of his opponents' shots, but it had a much larger effect because "they don't know which 5 percent it will be." In equilibrium, I got the commitment benefits of having students showing up at my door without actually having to pay very much to charity.

Still, the empiricist in me was puzzled that I was getting off so cheap. If anything, my having offered the "charitable bribe" seemed to dampen students' interest in running with me. The neoclassical economist in me couldn't understand why a charitable bribe would reduce demand. In retrospect, I now think that part of the problem might have resided in what behavioralists call "value ambiguity." In contrast to Tom Sawyer, who turned whitewashing a fence into a desirable opportunity, I had unwittingly turned "running with Ian" into a burden.

But an even larger problem with my charitable-bribe scheme may have been with my neoclassical faith in the dogma that you'll always get more of something when you offer a higher payment. Uri Gneezy, a business professor at the University of California at San Diego, has found in the lab that bribing people to work harder can backfire. He paid 160 students at the University of Haifa 60 NIS (Israeli new shekels, roughly $18 in today's U.S. dollars) for answering 50 questions taken from an IQ test. The control group was simply asked to answer as many questions as they could, while another randomly selected group was offered about 3 cents (10 NIS cents) for each question answered correctly. On average, students without the incentive answered 28.4 questions right. But students who could actually earn extra money for right answers averaged only 23.1 correct responses. Offering the extrinsic reward undermined the students' intrinsic motivation to answer questions correctly. And the pittance of 3 cents for not making a mistake seemed to insult some of the students—twice as many students responded to that bribe by answering all of their questions incorrectly.

The good news is that more substantial carrots—bribes of 1 shekel (about 30 cents) per correct answer—spurred more effort, as randomly selected students in this group answered an average of 34.7 (out of 50) questions correctly. Uri's title for the study nicely summed up his results: "Pay Enough or Don't Pay at All." The bad news is that studies of this kind suggest that for carrots to be effective, carrot contracts have to be fairly substantial (i.e., expensive) in order to work.

One more reason to prefer sticks—the incentive that keeps on giving—is that it costs you nothing as long as the goal is met. But even sticks

can produce perverse results. Uri was also behind a parallel study of ten
Israeli day-care centers that examined the impact of imposing fines for
picking up your kid late. Generally, the day-care centers in the study averaged about 8 late pickups a week. After four weeks of observation, six
of the ten centers were randomly selected, and these centers' personnel
announced to parents that those who were more than ten minutes late in
picking up their kids would be fined about $3, and that this amount
would be added to their monthly bill. It's Econ 101 that the fines should
have reduced the number of late pickups. But look at figure 6 to find out
what happened.

Figure 6. How Fines Increased Tardy Pickups at Day-Care Centers

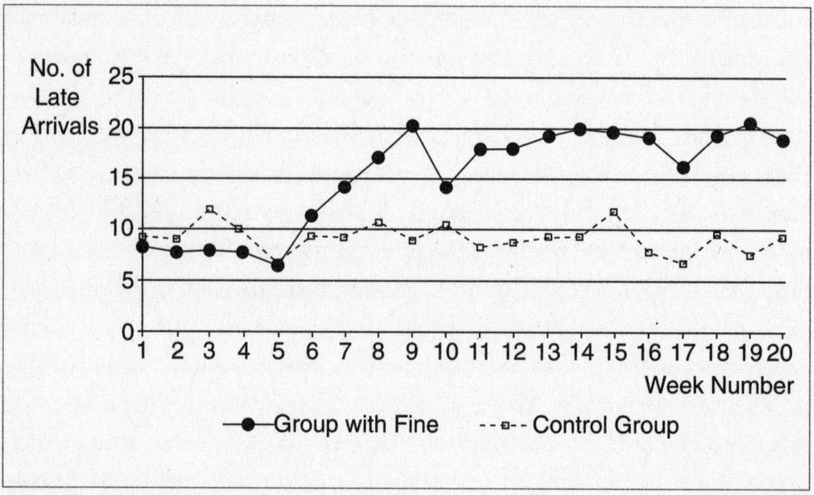

Source: Uri Gneezy and Aldo Rustichini, "Pay Enough or Don't Pay at All," *Quarterly Journal of Economics* 1156
(2000): 791, 804.

The number of late arrivals skyrocketed—more than doubling the tardy
pickups within just a very few weeks.

Uri's day-care study is a direct challenge to the earlier claim that
"losses loom large." Here, adding on a financial fine didn't take the choice
off the table—it seemed to make the option of coming late more available.
What gives? Parents who were fined didn't need to feel as guilty about
being late as long as they were willing to pay the "price." Uri and a cadre

of other behavioralists are beginning to think that financial incentives can backfire when they interfere with intrinsic motivation or social norms for doing the honorable thing. Before the fines, you were a "bad parent" if you showed up late to the day-care center, forcing some of the caregivers to wait. But after the center started imposing fines, parents felt that they had more of a right to come late as long as they paid the fine.

Uri's studies and even the failure of my charitable running bribe show that people can respond perversely to incentives. When you raise the price of an activity, you don't always get less of it. When you increase a subsidy for an activity, you don't always get more. Classical economics has trouble explaining these results because it ignores the social context. Incentives can have a perverse effect because they can also affect other people. This chapter is centrally about why other people matter in whether you will stick to your goals. What they think about you (or what you think they might think) can play a huge role in your likelihood of success.

One response to the day-care problem is to limit the use of carrots and sticks to contexts where there are not strong independent norms for good behavior. For example, use cash incentives only when social norms are weak or ineffective. The day-care centers might have been more successful if they had taken a page out of Zappos' original playbook. They might have still used cash but tried a kind of anti-incentive. Strange as it may sound, imagine that the center announced that whenever a parent came late they would force one of the (poorer) caregivers to pay money to the offending parent. The very injustice of the incentive scheme might reinforce the anti-tardiness norm. An analogous anti-incentive has at least been proposed in a literary day-care setting. In *Little Men* (Louisa May Alcott's sequel to *Little Women*), the schoolteacher Mr. Bhaer proposes a punishment for lying that he hopes will be harder for students to treat merely as a price. He tells Nat, "When you tell a lie I will not punish you, but you shall punish me. . . . You shall ferule me in the good old-fashioned way." And, in fact, Mr. Bhaer later forces a distraught Nat to give him "six good strokes." In the story, Nat never lies again. Marine drill instructors are masters at this kind of incentive training: an entire platoon of recruits may have to hit the deck because a single member fails to lace his or her boots correctly. Sometimes punishing the innocent changes behavior. Or, as we're about to see, you might want to promise to reward the unworthy if you fail.

"I DONATED TO NARAL"

Rob Harrison is one of the most beloved teachers at Yale Law School. He has improved the writing and emotional outlook of generations of our students. He is the kind of guy who unabashedly ends his emails with "Love, Rob." He is staggeringly kind. So it came as a bit of a shock when he told me that he had used unforgiving commitment contracts to help students overcome writer's block. For more than a decade, students have given him checks of up to $10,000, signed and made out to various charities, and authorized Rob to mail the checks if they failed to turn in a paper to the course professor by a specified date.

What's really interesting is the charities that the students chose to potentially fund. For the first five years that Rob provided this service, the procrastinators made the checks payable to charities they liked. But about five years ago, a student suggested that making the checks out to charities they *didn't* like would be an even more effective incentive. The great news is that Rob has never had to mail one of these commitment checks. This is an amazing result—particularly because Rob offers the contracts only to students who have already demonstrated a deep psychological inability to put pen to paper (or, nowadays, fingers to keyboard).

At the end of the last chapter, I told the story of a friend who put at risk her $5,000 annual contribution to her college. If she failed to make her goal weight, she wouldn't be able to give to her beloved alma mater. But people who want to "put their money where their mouth is" with a stick contract can supercharge their motivation by committing to send any forfeiture to an anti-charity, an organization they actively oppose. Committing to forfeit money to an anti-charity can reinforce the idea that the money at risk is not merely a price that can be paid if the commitment becomes inconvenient. Committing to forfeit money to an organization you despise is a commitment to take the future choice off the table. Far from disrupting the independent norm to do the right thing, anti-charities can help tailor incentives to promote or reinforce these independent norms for action. The day-care centers' fines would have been more effective if parents had known that any forfeited money would go to some cause they strongly disliked; it's hard to put a price on funding a political party you detest. Anti-charities can again make small financial losses loom large—even those that would otherwise be mere prices. *Lit-*

tle Men punishments and anti-charity rewards work because we're social creatures. We care about how our actions affect others. With anti-charities, the threat of guilt and the loss of cold, hard cash is a powerful combination.

University of Chicago law professor Eric Posner sees only mischief in the idea of anti-charities:

> What stickK needs to do is donate your money to an anti-charity, a group that almost no one approves of: the government of Sudan, say, or the tobacco industry. Perhaps if you know that breaking your diet makes you complicit in genocide, you will resist that slice of chocolate covered cheesecake. Let's hope that the Sudanese government does not realize that it can cut out the middleman, and make money from obese Americans directly, by agreeing to take their money if they fall off their diets. Indeed, if [stickK's] business model is sound, we can imagine a future emporium of self-control entrepreneurs, with Sudan, North Korea, the tobacco industry, baby seal hunters, and pedophiles all vying for the business of the overweight, the underdeveloped, and other sufferers from weakness of will.

But Posner forgets that there is a limit to the kinds of groups people are willing to even potentially support. Few people will be able to bring themselves to sign a contract where their money is given to baby-seal hunters or the Nazi Party. The choice of anti-charities is another dimension where we can see the tension between what would be most effective and what people are actually willing to agree to. Here, the participation constraint is likely to keep the most extreme anti-charities merely as hypotheticals for overwrought bloggers.

The idea of anti-charities is value-neutral. To work, all a user needs are deep-seated allegiances or preferences. For some, forfeiting funds to the Yankees would inflict the requisite pain. For others, it might be giving (extra) money to the IRS. And stickK.com is here to help. The website allows you in advance to designate as a recipient of your funds any "friend or foe" or any charity or anti-charity you want. If you care about gun control, you can have your money go to either the NRA Foundation or the Educational Fund to Stop Gun Violence. Depending on your political persuasion, you could choose to have your stake go to the George

W. Bush or the William J. Clinton presidential libraries. I personally support gay rights (and have even written an academic book suggesting what straight allies can do to support the cause). But stickK doesn't take a stand on this issue. It lets users direct their money to groups that promote or oppose same-sex marriage.

In the fall of 2008, I came across this post from the "Restrained Radical" blog, entitled "I Donated to NARAL"—the National Abortion Rights Action League:

> I made a $10 donation to NARAL Pro-Choice America this week. OK, it was semi-automatic. I made a stickK commitment to work out at least five times a week. If I fail in any given week, my credit card is charged $10 and donated to the anti-charity I chose. I was very busy and only worked out four times last week. I could have lied and reported that I worked out more than I actually did. Instead, I decided to let the donation go through but also to make a matching donation to a pro-life group. I have to find a just cause now.

This short post underscores the binding nature of even self-refereed commitment contracts. Some people are not willing to lie (and break their promise to report truthfully), even to a faceless Internet site. But it also shows the limits of anti-charities in eliminating future choice. When the Restrained Radical originally entered into the exercise commitment with NARAL as the designated anti-charity, the thought of failing was probably inconceivable. You can't promote the killing of baby seals on stickK, but people in the pro-life movement can support what they may view as the (legal) killing of innocent babies. Notwithstanding the intensity of this before-the-act motivation, when life got busy, the Restrained Radical found a way to monetize the unthinkable. Giving an offsetting amount to a pro-life (anti-choice) group would counterbalance the offending impact of giving to a pro-choice (pro-abortion) group. The price of failure is double, because effectively $20 instead of $10 is now on the line. But with the offset, failing to go to the gym is now a mere price. One can imagine more comprehensive charity commitments—ones that would force users to cut their contributions to preferred charities and increase their contributions to dispreferred charities. But the larger point is to see that our future selves will be devilishly inventive in trying to figure out ways to rationalize forfeits that we couldn't initially imagine ever paying.

THE NORM TO CONFORM

As the Regents' Professor of Psychology at the University of Arizona, Robert Cialdini wasn't focused on how to win friends. But he has written the definitive books on how to influence people. A few years ago, he nearly jumped out of his seat when he heard a graduate student tell a story of stopping with his fiancée at Arizona's Petrified Forest National Park on his way to Tempe to begin work with Cialdini. Near the entrance, the couple encountered a large sign that read, "Your heritage is being vandalized every day by theft losses of petrified wood of 14 tons a year, mostly a small piece at a time." Cialdini told me, "Before the student finished reading the sign, he heard his fiancée say, 'We had better get ours now.'" Her comment was jarring because it was so out of character. The student described her "as the single most honest person he had ever met in his life," Cialdini said. "She never borrowed a paper clip that she hadn't returned."

The fiancée's line "We had better get ours now" excited Cialdini because he's always on the lookout for big effects. "When something," he said, "causes honest people to be cheats, be criminals, and in this instance to deplete a national treasure in the process, there had to be a powerful effect there." Cialdini had a hunch that the sign was backfiring because it was implicitly sending the message that many, if not most visitors were pilfering. People have such a strong inclination to conform their behavior to what they think other people are doing that it can overcome their urge not to cheat or steal.

But as a good behavioralist, Cialdini wasn't satisfied with theorizing. He went out and ran an experiment. Cialdini convinced park officials to let him place secretly marked pieces of petrified wood along pathways visitors would travel. Over five consecutive weekends, he varied the signs at the entrance to each path. What's truly amazing is that putting up no sign at all did a better job of discouraging the taking of wood than posting a sign that emphasized the pilfering problem. "They were undermining their own goals by inadvertently communicating that everybody steals, by normalizing theft. And the consequence was that theft went up almost threefold," Cialdini said. "Theirs was not a crime-prevention strategy; it was a crime-promotion strategy."

The big idea here is that people want to be like others. If they hear that most other people like them are doing something, they will want to do the same thing. There is a powerful pull to conform. Want to get hotel guests to forgo daily towel cleaning? Don't give them a card telling them that reuse will help the environment; give them a card telling them that most other guests reuse their towels. Want even more people to participate? Tell them that most people using that very room opted for reuse. This isn't just idle speculation. Cialdini did this study, too. New cards were printed up that said:

JOIN YOUR FELLOW GUESTS IN HELPING TO SAVE THE ENVIRONMENT

Almost 75% of guests who are asked to participate in our new resource savings program do help by using their towels more than once. You can join your fellow guests to help save the environment by reusing your towels during your stay.

Guests who saw that message were 26 percent more likely to reuse their towels than those who saw the basic environmental-protection message; those who learned that most others who had stayed in their room had reused towels were 33 percent more likely to reuse.

I've been devoting a major chunk of my research simply to applying Cialdini's peer-pressure idea to different contexts. In the fall of 2009, I taught a seminar titled Incentives and Commitments. The class was in part a commitment device to force me to articulate a bunch of the ideas in this book. But I also sent the students out into the world to do field experiments. Stephanie Tang emailed graduate students about how often they were taking advantage of a free-lunch mentoring program at the undergraduate dining halls. Just as Cialdini would suspect, graduate students who were told that they were lagging behind their peers in attendance showed a 19-percentage-point increase in the following semester relative to a randomly selected control group that received no email.

Ryan Sakoda and I ventured slightly further afield. We wanted to see whether Cialdini's simple idea could help improve teaching. Virtually every college and university has student evaluations of teaching quality, but we have never found a college that gives effective feedback about how well professors teach relative to their colleagues. Some schools will tell a professor how her score compares to the average score. But we've never heard of a school that tells a teacher her percentile ranking. Ryan and I decided to correct this omission. We designed a "bad teacher" test for which we downloaded the publicly available teaching-evaluation data from three different universities and calculated the percentile rankings for particular courses. Then we randomly selected some of the teachers with the worst evaluations and sent them an email from my very own email account (ian.ayres@yale.edu) with a message about how well they were doing relative to their peers. For example, an email would explain that as part of my research on the effectiveness of teaching evaluations, I had analyzed the evaluations for a particular course—say, Introduction to Forestry. It would then in bold type reveal the professor's percentile ranking:

> Your student evaluation rating for overall quality in this course during the past two academic years was in the bottom 7% of all courses taught at [your university] and in the bottom 5% of all courses taught within the School of Forestry.

The simple idea was to come back the next semester and see whether the professors who received these emails earned improved evaluations relative to those of similar professors who by chance didn't receive the emails. Our hypothesis, by the way, was that there would be a general improvement toward the average score in both groups. Some of the poorly evaluated professors likely just had an off year because of some personal distraction and would do better regardless of whether they received an email from me. But the crucial question was whether the professors who received my email bounced back stronger than those who didn't. If Cialdini is right, you should feel a powerful urge to change when you learn for the first time that 95 percent of your colleagues are doing better than you.

The results aren't in yet. And this study may not, in fact, see the light of day. After I sent the emails, I got a variety of responses. Some professors thanked me for providing them with useful information. But more were upset. Some professors contested our calculations. Others claimed that they had already been given this information. Many wondered why it was any of my business to send them an email out of the blue commenting on the quality of their teaching. In response to the complaint of one school administrator, Yale's Human Subjects Committee has for the moment shut down the project. Following federal and university rules, I had sought and gained the committee's permission to conduct this experiment on the professors. But after an administrator at one of the universities complained, Yale's committee reversed itself and instructed me to break off analyzing any of the data from the three schools. Even though committees at Yale and other schools routinely approve similar studies—like Tang's emails to graduate students and Cialdini's petrified-forest and towel tests—something about testing professors hit a nerve. The norm to conform is powerful even when it comes to what kind of testing is appropriate. When it's testing on someone like you (that is, another professor), suddenly the need for informed consent becomes paramount.

Luckily, I'm still allowed to analyze two other experiments that apply Cialdini's idea outside of the academic groves. Working with Sophie Raseman and Alice Shih, I've tested whether peer information can help people reduce their energy consumption. With the help of OPOWER, a cutting-edge energy messaging company, we analyzed randomized experiments on more than 150,000 homes in Sacramento and the Puget Sound area.

The basic problem with a traditional energy bill is that it fails to tell you whether you're wasting energy. Sure, it tells you how much you used last month and lets you compare that with how much you used in the past. But it fails the Cialdini test. It doesn't let you know how you're doing relative to people like you—in similar-sized homes, facing similar weather conditions. OPOWER has redesigned the energy bill to give you just this kind of information. Figure 7 shows how much more information you'd get if you received OPOWER's "Neighborhood Comparison," which was the type used in the Sacramento mailings.

The smiling emoticon (☺) quickly tells you that you're doing "good" but not "great." Compared to neighbors in similar-sized (1,635-square-foot) houses, you are consuming less than the average. But 20 percent of your nearby neighbors are using a lot less energy and, as a result, are saving over five hundred bucks a year. You'd have to wonder what they're doing differently.

What's the impact of receiving monthly reports like this? OPOWER generously gave me a first crack at analyzing the data. This beautiful ran-

Figure 7. Energy Use Comparison

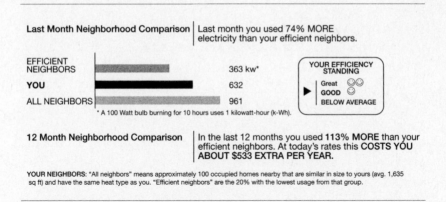

Last Month Neighborhood Comparison | Last month you used 74% MORE electricity than your efficient neighbors.

EFFICIENT NEIGHBORS — 363 kw*
YOU — 632
ALL NEIGHBORS — 961

* A 100 Watt bulb burning for 10 hours uses 1 kilowatt-hour (k-Wh).

YOUR EFFICIENCY STANDING

Great ☺☺
▶ GOOD ☺
BELOW AVERAGE

12 Month Neighborhood Comparison | In the last 12 months you used 113% MORE than your efficient neighbors. At today's rates this COSTS YOU ABOUT $533 EXTRA PER YEAR.

YOUR NEIGHBORS: "All neighbors" means approximately 100 occupied homes nearby that are similar in size to yours (avg. 1,635 sq ft) and have the same heat type as you. "Efficient neighbors" are the 20% with the lowest usage from that group.

Source: www.opower.com/Approach/TargetedMessaging.aspx.

domized control experiment suggests that randomly selected Sacramento households receiving the monthly report have reductions in their energy consumption of 2 percent sustained over a six-month period. A 2 percent improvement might sound like just a matter of chance. But look at figure 8, where we plotted the average difference in energy usage between the treatment group and the control group, to see what happened over time.

There was virtually no difference in usage between the two groups in the preexperiment period. But as soon as the first neighborhood comparisons started to be mailed (on a monthly basis) in April 2008, the "treated" families showed a marked (and statistically significant) drop in energy consumption.

What's more, we found that the comparison information had the biggest effect on the people who needed it most. Figure 9 shows the impact of the messages on households that in the past had used different amounts of energy.

At the far right of figure 8, you can see that households that were in the highest 10 percent of energy users on average reduced their energy consumption by more than 6 percent after they were informed about how much energy they were wasting. The energy hogs learned that they were consuming more energy than their peers and cut back. And the neighborhood comparisons accomplished this without inducing the conservationists to consume more. There's always the risk of a boomerang effect, where the better-than-average peers will revert to the

Figure 8. The Effect of the OPOWER Messages on Energy Usage

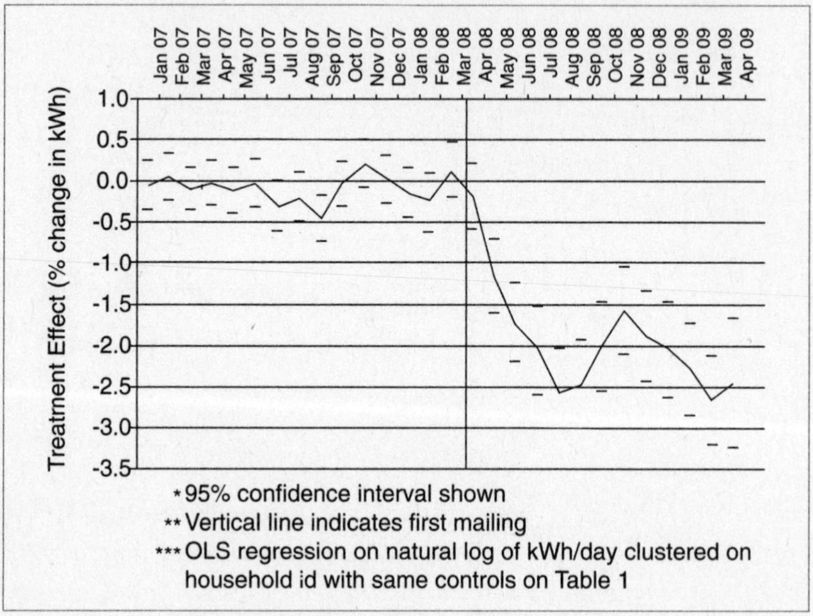

* 95% confidence interval shown
** Vertical line indicates first mailing
*** OLS regression on natural log of kWh/day clustered on household id with same controls on Table 1

Source: Ian Ayres, Sophie Raseman, and Alice Shih, "Evidence from Two Large Field Experiments That Peer Comparison Feedback Can Reduce Residential Energy Usage" (working paper, July 16, 2009).

more mediocre mean. That's what happened at the petrified forest when the honest fiancée learned that she was more virtuous than the average visitor. But the far left of the figure shows that the lowest-consuming households had no appreciable increase in energy use. These consumers didn't want to lose that smiley face. They liked the normative seal of approval. And OPOWER was doubly smart in giving the "efficient neighbor" comparison as a separate standard to try to reach. What started as cute studies about petrified wood and hotel towels has turned into an engine that can have lasting impacts on all kinds of individual behavior.

I'm in the midst of seeing whether peer comparisons can affect civil rights. Since 2003, I've been working with Equality Forum's Fortune 500 Project. We've written to the CEOs of Fortune 500 companies with a simple request: that they include sexual orientation in their company's nondiscrimination policy. We've had a tremendous response. Several of the companies that have adopted wording in response to our letters said that they'd never been asked. But a crucial aspect of our approach was inspired by Cialdini. Our letters also told the noncompliant companies

Figure 9. How the OPOWER Bill Affected Households with Different Energy Usage .

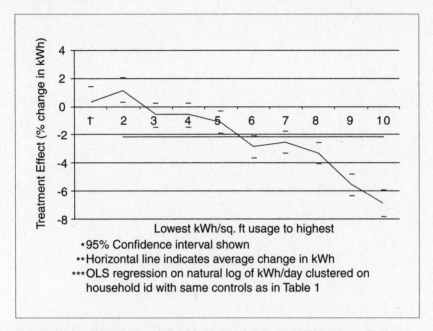

Source: Ian Ayres, Sophie Raseman, and Alice Shih, "Evidence from Two Large Field Experiments That Peer Comparison Feedback Can Reduce Residential Energy Usage" (working paper, July 16, 2009).

about how many of their peer companies have express policies against discrimination on the basis of sexual orientation; in 2008, the proportion of companies that extend such protection had grown to a whopping 94.6 percent. But we provided more localized comparisons, as well. Our letters also specified the proportions of companies in the same industry and companies with headquarters in the same state that had adopted the nondiscrimination clause. For example, after we told Assurant that out of fifty-five Fortune 500 firms headquartered in New York, it was the only one without an express sexual-orientation protection, it quickly changed its policy. Like people, corporations don't want to be aberrant.

Every year, the composition of the Fortune 500 changes somewhat, as the revenues of individual companies rise and fall, and so every year I join Equality Forum in sending out a new batch of letters to the few recalcitrants. (Equality Forum maintains a cool website where you can easily find the holdouts by state and industry. Hint: think Texas tea.) But another student from my seminar, William Rinner, is helping me run a

Cialdini-like experiment (with human subjects' approval!) in which we have identified noncompliant firms among the five hundred next-largest firms in America—the bottom half of the Fortune 1000. Some of these firms are just being asked to do the right thing; others receive a letter that adds peer comparisons. It's too soon to report the results, but don't bet against the norm to conform.

A big part of the urge to behave like others is the simple idea that there is safety in numbers. You can take advantage of the wisdom of the crowds; it's good to know where the locals eat. But another part of the norm to conform stems from a deep-seated desire to be liked. We mirror the body language of others because we care about what other people think of us. The desire to keep up with the Joneses can be a powerful motivator.

NAGGING AND BRAGGING

Barry Nalebuff is obsessed with real-world game theory. When I first started hanging out with him, I was slightly annoyed that he kept posing math and logic puzzles for me to solve. Was he testing me, or what? Turns out that Barry tests everybody, including himself. Barry, who looks a bit like the actor Paul Giamatti (with a bit more hair), is constantly trying to figure out whether there is a better way to organize the world. I should disclose that Barry and I are almost joined at the hip as writers— with two books, a long-running column in *Forbes*, and more than sixty publications coauthored.

But it is something Barry did back in 1993 that underscores how much people care about how they are viewed by others. Barry was teaching a business school course on game theory, and as a young professor he decided to try something different. "The idea was to help students appreciate games that you played against your future self," Barry told me, "and I thought I would demonstrate that to them the first day of the class. And so I came in with a Thinner scale, one of those small scales that you step on twice and it resets, and said, 'I am a little pudgy and want to lose fifteen pounds this semester and I am pretty confident that if I tell you this and commit to teach my last class in a Speedo if I haven't, that I will lose the weight.'" He got on the scale then and there and promised to do so again during the last class. He then added, "I don't think it is fair that I be the only one who has the opportunity to take advantage of this contract, and so any student who wants to join in is welcome to do the same thing." About a dozen of the sixty students took him up on the offer.

The idea that the threat of humiliation can motivate better behavior has a long history in game theory. In 1971 the Nobel Prize–winning economist Thomas Schelling wrote about a Denver addiction clinic that used "self-blackmail as part of its therapy":

> The patient may write a self-incriminating letter that is placed in a safe, to be delivered to the addressee if the patient, who is tested on a random schedule, is found to have used cocaine. An example would be a physician who writes to the State Board of Medical Examiners confessing that he has violated state law and professional ethics in the illicit use of cocaine and deserves to lose his license to practice medicine. It is handled quite formally and contractually, and serves not only as a powerful deterrent but as a ceremonial expression of determination.

The opportunity to engage in contingent self-extortion can increase your resolve.

But Barry's humiliation game ended almost before it began. Barry's boss at the time, the dean of Yale's School of Management, Paul MacAvoy, was none too pleased. MacAvoy was a rather Brahmin, blue-blood type who routinely dressed in a tweed jacket, J. Press shirt, and tie. "MacAvoy explains to me, the young whippersnapper that I am, that it really would not be proper decorum to teach a class at Yale University wearing nothing more than a Speedo," Barry related to me. "To which I replied, 'I couldn't agree with you more.' I had absolutely no intention whatsoever of doing that. If I had any doubts whatsoever at my chance of success, I wouldn't have done this." MacAvoy ultimately relented. Indeed, Barry feels his boss's displeasure actually added an extra incentive.

As the semester proceeded, the class closely watched the progress of Barry and the student participants at interim weigh-ins. Some of Barry's students tried to sabotage his efforts by giving him Twinkies and ice cream and the like. The class wanted to know whether a fancy professor would really end up next to nude. And it came down to the wire. "I am pretty sure," Barry said, "during the last week I was within two pounds of my goal. And then, that last week, I probably overdid it in terms of fasting and running so that there wouldn't be any extra doubts, and I think I ended up losing about sixteen and a half pounds. What was more remarkable in my view was that [of] those students that joined in, all but one of them lost the requisite amount of weight."

The self-blackmail worked so well, Barry figured it was ready for prime time. In 2006, after Tom Schelling had won the Nobel Prize, in part for his work on commitment devices, Barry designed a humiliation experiment for ABC's *Primetime*. He found five perfectly sane but overweight people and invited them to a photographer's studio. Four women and one man "bared their souls and plenty more in front of the cameras." It was just like the *Sports Illustrated* swimsuit issue—except that the models weren't bikini-ready.

If things worked out as planned, the pictures would never be shown. ABC promised to destroy the photos if the participants lost fifteen pounds over the next two months. If they didn't lose the weight, ABC promised to show them on national TV. "The key to making it successful," Barry said, "was finding people who were not exhibitionists, who truly would be embarrassed by showing themselves sticking out of a bikini or Speedo on prime-time television. And so, therefore, we needed to find normal people, as opposed to the one aspiring actor or whatever." The male participant was actually worried that he wouldn't be sufficiently self-conscious about having his overweight picture shown on TV. "And so what we did there is we had his wife also be photographed, not just him," Barry remembers. "And the deal was that her photographs would be shown along with his if he didn't lose the weight. So, as he put it, he needed to lose the weight or lose his wife."

The participants knew that being overweight carried serious health risks. Yet, as Barry wrote in *The Art of Strategy*, "That wasn't enough to scare [them] into action." Cindy Nacson-Schecter explained that what she "feared more than anything was the possibility that her ex-boyfriend would see her hanging out of a bikini on national TV." And she knew he would be watching the show, because her best friend had already told him to tune in.

The threat of national exposure was a powerful motivator. "We didn't do anything else," Barry told me. "We didn't give them trainers. We didn't give them fitness coaches, nutrition coaches. But we did email them electronic versions of the photographs. So that they could see what they looked like. And it was a fashion photographer, so they were really high-quality, wonderful photographs. But they were also rather incriminating."

Two months later, the humiliation diet was a grand success. Cindy had lost seventeen pounds. Ray was down twenty-two pounds (and spared his spouse from exposure). As in Barry's class experiment, all but one of the dieters made their goal. But even the failure had succeeded in losing thirteen pounds and was down three dress sizes. "In the end," Barry said, "the host of the show gave her back the photographs and didn't show them on TV, sort of as a kind gesture."

While the potential humiliation of being exposed on ABC would do the trick for most of us, this isn't a practical solution. But we can come close by telling our friends and family about our goals and asking them to give us grief if we fail. My childhood hero, Jim Ryun, famously stood up as a high school freshman in Kansas and promised to break the four-minute mile before he graduated. After I bragged about my weight-loss success on the *New York Times'* Freakonomics blog, Justin Wolfers (an economics wunderkind from the University of Pennsylvania) responded:

> There is a much cheaper way to commit your future self to some targets, since the fear of failure, public ridicule, and embarrassment can be harnessed (for free!) to help solve our own commitment problems. Let me explain with an example.
>
> I'm going to publicly declare my major fitness goal on this blog, and rely on this blog's readers to ridicule me if I fail. So, here goes: this summer, I'm going to be visiting the IIES at Stockholm University, and on the last day of my visit, I'm planning on running the Stockholm Marathon. And I hope that you, dear reader, will keep me honest. It would be embarrassing to fail publicly, and I suspect it would be embarrassing enough that today's public statement of my running goals will keep my future self pretty darn motivated.

Justin followed through (and finished the marathon in Stockholm in four hours and fifteen minutes). You don't even have to blog at *The New York Times* to make this work. You can stand up and tell your friends, your coworkers, or even strangers about your resolutions and ask them to ask you about your progress. But don't forget Cialdini when deciding whom to tell. We have a particular desire to make our behavior conform to that of other people like us. So if you want to quit smoking, you'd probably be better off telling nonsmokers of your plan to quit than smokers (if you plan to lose weight, tell skinny people).

THE HUMILIATION CARROT

The potential to humiliate someone else can also be harnessed as an incentive. In New Zealand, Craig Forsman went further than Nalebuff. He appeared in the buff—but it wasn't because he failed. It was because his friend succeeded. Forsman agreed to run naked through the middle of Tauranga if his friend (and boss!) Peter Harford quit smoking for a month. Humiliation incentives might motivate some people because of a kind of schadenfreude. But you have to worry about "victims" who really want the excuse to expose themselves and be the center of attention. I know: I once offered to shave my head if the response rate on a student survey exceeded 75 percent.

Or you might follow my example and actually hire a professional nagger. That's right. At the beginning of 2007, I posted an ad in the Yale student employment office saying that I wanted to hire someone who was willing to nag me if I failed to meet specific goals. The nagger's job was simple. Once a day, the nagger would log in to a Google spreadsheet document I'd created and check whether I was keeping each one of five different commitments. I had made resolutions to read thirty pages, write eight hundred words, and exercise for twenty minutes every day. I'd also committed to losing weight and lifting weights every other day. The nagger's job wasn't to make sure that I was in perfect compliance every single day, but if she saw that I was failing for two or three days in a row (including failure to make entries on the spreadsheet), she was supposed to intervene. If I started to slip, her instructions were first to email me. If the problem persisted, she was instructed to call and even come by my house to, well, nag me into compliance. A by-product of this nagging regime is that I was forced for the first time to keep a daily calendar of a few problematic areas of my life.

Nagging has worked for me like gangbusters. I committed to exercising more, and that first year I swam over one hundred miles. Even more amazingly, for the first time in my life I can do ten unassisted pull-ups. (High school classmates who correctly remember me as an emaciated cross-country runner will not believe that this is possible.) I committed to reading more and have had the pleasure of reading more than forty novels. And most important, I committed to writing more and was able to submit the manuscript of my last book, *Super Crunchers,* on time.

Or, to be more accurate, the *threat of nagging* has worked for me. You see, the nagger had to email me only about half a dozen times during the year (and the intervention never needed to proceed to a telephone call or the dreaded home visit). I find it deeply embarrassing to have a student call me on the carpet—and so was well motivated to make sure that it didn't happen. One of the coolest things about commitment contracts is that they can be both effective and cheap. Nagging wasn't free, but paying someone to log on once a day and send a total of five emails over the course of the year was incredibly inexpensive. This year, I'm experimenting with an even cheaper version where the nagger at least initially has to check my online diary only once a week.

TOO SKINNY?

My success at increasing my upper body strength and lowering my BMI from 25.3 (overweight) to 22.2 (normal) was not universally greeted with praise. Two different members of my family repeatedly warned me that I looked too skinny. Even though I was objectively in the best and strongest shape of my working life, loved ones (who genuinely have my best interests at heart) felt comfortable giving me unhealthful advice. There is a great asymmetry with regard to weight. Members of my family, who would never say to a relative that they weigh too much, were willing to tell me that I weigh too little. Our nation's obesity epidemic has reached a point where a "normal" BMI now looks abnormally skinny. The upshot is that in trying to stick with your commitments, you have to worry about "anti-supporters" who react to your successes as failures.

In Japan, a new website lets you sign up for a virtual wife who will send you nagging reminders to eat your vegetables. Middle-aged salarymen can choose a virtual nurse, security analyst, manicurist, or maid, and can even pick her name. If you fail in meeting your goal, the tone of her emails morphs from gentle to increasingly bossy, but in the "voice" appropriate to the character. See if you can match the pictures and profiles in figure 10 to the occupation.

Figure 10. Nagging Virtual Wives

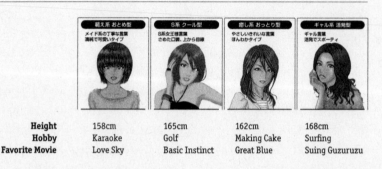

	萌え系 おとめ型	S系 クール型	癒し系 おっとり型	ギャル系 活発型
	メイド系の丁寧な言葉 純粋で可愛いタイプ	S系女王様言葉 さめた口調、上から目線	やさしいきれいな言葉 ほんわかタイプ	ギャル言葉 活発でスポーティ
Height	158cm	165cm	162cm	168cm
Hobby	Karaoke	Golf	Making Cake	Surfing
Favorite Movie	Love Sky	Basic Instinct	Great Blue	Suing Guzuruzu

Source: www.metaboinfo.com/okusama/.

ON YOUR HONOR

To be honest, I find the icky sexism of the website a bit creepy. (Is it really important that the maid be short?) But as a commitment device, the site has the independent problem that it cannot tell for sure whether any of its clients is really making progress. Your virtual manicurist asks you for your weight. But what's to keep you from lying?

For some people, the answer is "nothing." If you enter into an "on your honor" contract at stickK and then tell the website that you have succeeded in keeping your commitment, you're not going to forfeit any money—regardless of whether or not you actually kept your goal. But other people have trouble lying. The Restrained Radical was even willing to forfeit money to what he or she thinks is an immoral cause rather than lie. Some people have trouble breaking promises to themselves. George Ainslie, the pigeon researcher from chapter 1, told me that he sometimes promises himself that he will not eat another dessert for the rest of the week, and seals the bargain by throwing a coin into the air and catching it. So far, he's never broken one of these "promised myself" commitments. But he's careful not to ask too much, because he realizes that the device can bear only so much stress, and he doesn't want to lose the magic.

"On your honor" contracts at stickK, however, are more than promises to yourself. You are making a commitment with a legal entity. And when it comes time to report on your success or failure, some people—even in this day and age—have difficulty lying even when no one is going to be able to call them on it. In fact, an upside of monetary forfeitures is that people have a harder time being dishonest when it comes to money. Many employees think nothing of taking a pen home from work but would never take a quarter from the cash register.

Nina Mazar, a marketing professor at the University of Toronto, ran a randomized experiment concerning honesty and incentives. At a student cafeteria at lunchtime, she offered students the chance to participate in a five-minute experiment where they'd be asked to solve twenty math problems—and be paid fifty cents for each correct answer. Since Mazar wanted to study honesty, she designed the experiment to be an "on your honor" incentive. When students were done, they were told to grade their own answers, tear up their work sheet, and "stuff the scraps into their pockets or backpacks, and simply tell the experimenter their score in exchange for payment." Half the students were then paid directly in cash. But, at random, half the students were paid instead with a token, which they could then redeem for cash simply by walking twelve feet across the room to a second experimenter. Because of randomization, it is unlikely that the students in these two groups differed in their ability to answer the questions. But sure enough, there was a significant difference in the number of questions the students reported getting right. Students who were paid directly in cash on average said they answered 6.2 questions correctly, while those who were paid first in tokens said they answered 9.4 problems correctly. It's not just the Restrained Radical; a lot of people have a harder time lying when they're thinking about money.

Harder doesn't mean, however, that it is impossible to lie about money. Mazar randomly assigned other students to a third group, where the experimenters directly graded how many answers were correct. And these students on average answered only 3.5 questions correctly. Sometimes we need something more than our disinclination to lie to keep our futures selves committed. As Ronald Reagan used to tell Mikhail Gorbachev: "*Doveryai, no proveryai.*" (Russian: Доверяй, но проверяй).

Trust, but verify. Instead of self-verification, it will often be more effective to designate someone else to referee whether or not you in fact succeed in sticking with it. But as with so many other commitment issues, the choice of referee can be fraught with complications. First off, you have to trust that the referee will be fair. If you designate a stranger to be your referee on a commitment contract where you have put money at risk, that referee has the final say in deciding whether you will forfeit money. You better be sure that your referee doesn't dislike you.

An even bigger problem is that you need to trust that the referee will be willing to follow through and forfeit your money if you do fail. Too often referees wimp out. Remember Alex Moore, the sneezer from the Introduction? His commitment to show up to work on time didn't work out so well. Or, to put it more bluntly, he failed to keep his commitment. But his girlfriend, who knew he was sleeping late, let him off the hook. They ended up lying. A financial stick won't do as much work if you suspect that your ref is going to let you slide. So when you are setting up a commitment contract, you shouldn't designate either an enemy or a soft-hearted friend to be your referee.

Who's left? Well, economist friends are pretty good. Barry Nalebuff is a great referee. As a game theorist, Barry definitely gets the importance of follow-through. And he's thought long and hard about how to structure a threat so that it really is credible. When Barry opened up the Speedo bet to students in his game-theory class, he wondered how he was going to enforce the promise. He knew he'd face a class revolt if he breached his promise to parade. But what would he do if one of his students failed to lose the weight and then refused to go on display before the class? Should Barry fail the student? (Luckily, the one student who failed to lose the weight wore the swimsuit without complaining.) Now, thinking ahead, Barry tries to avoid being put in tough situations like that.

So when *Primetime* called, Barry said, "We got a little smarter. We photographed them *before* they lost the weight. We started with the upper hand in terms of actually having the photographs." If the people failed to lose weight, there wasn't a question of whether they would be willing to go on display, because the producers already had the incriminating pictures in their possession. But as it turns out, even that wasn't enough. When one of the participants failed, the show's host didn't want to be seen as a jerk, and wasn't willing to follow through and show the woman's bikini photo on the air.

Barry was disappointed. *Primetime* was blowing its (and, indirectly, Barry's) credibility to threaten unpleasant consequences in the future. So when *Primetime* called again and asked to do a second show on bikini bets, Barry wasn't willing to take the producers at their word. For the second broadcast, Barry involved a professional baseball club, Connecticut's own Bridgeport Bluefish. Not only did the weight-loss participants take pictures in advance; the plan was to show them on the Jumbotron during a game. "We had them doing yoga poses over home plate," Barry said. They were not only in skimpy swimsuits but they were contorted in poses like the downward-facing dog and the salutation seal. "They were some of the funniest pictures you would ever see in your lifetime."

The key was to get other people to act as enforcers. "To the extent that the team had advertised that these pictures would be coming and they had this public commitment," Barry said, "it would be harder for them to back out." Barry even tried to get the producers and the show's host to commit to having their own pictures taken in advance, so that he could show these if *Primetime* again failed to follow through on its commitment. But they refused. Barry should have figured something was up. In the end, one of the weight-loss bettors failed and again *Primetime* reneged. Even the best-laid plans of one of the world's great game theorists weren't enough. "In the end, the Bluefish never really did the advertising to create the public commitment and so . . . [the producers] wimped out," Barry reported. "It's now at the point that I couldn't do the show with *Primetime* again because there is a question of whether they could ever make that commitment credible."

Barry was upset because *Primetime* was screwing with his own reputation—not once, but twice. A reputation for following through on your promises is hard to establish and even harder to reestablish once you've been shown to be the kind of referee who is not willing to pull the trigger on a forfeiture. One of the coolest things about eBay is that it has created a wonderfully helpful reputation market for the quality of both buyers and sellers. You can trust that a particular seller is actually going to send you a mint-condition George Brett baseball card because the seller has a positive feedback rating of 99 percent.

A similar feedback system could work for referees. I imagine a day when an army of stickK referees would post ads on craigslist offering, for a fee, to referee other people's commitment contracts. Refs could make sure that you take off the weight by having you get on the scale. They could make sure that you exercise by actually exercising with you. Indeed, I might have been wiser to offer to referee my students' own running commitments if they gave money to charity—rather than trying to bribe them to help me commit to run. Framing my participation as reffing metaphorically takes me out of the picture and turns me into a mere observer. Or imagine the reffing possibility for your local YMCA. By serving as an exercise referee, the Y (or Bally or Planet Fitness) could drive new members to join—because the ref would be assessing whether you were using the facility. All of these possibilities would work better if clients and others could complain about refs who failed to follow through.

I've taken a baby step toward my dream of a vibrant referee network. If you go on stickK, you can easily find a user named NewHavenRef whose profile says, "For a fee, I am happy to serve as a referee on your commitment contract for anyone living near New Haven, CT." From tiny seeds . . .

THE DIRTY MAN COMPETITION

In the movie *Knocked Up*, five roommates entered into a strange commitment contract not to shave or cut their beards for a year. Anyone who shaved had to pay the rent for those who didn't. When the movie opens, Martin is the last man standing. All the other roommates have succumbed to the temptation of the blade. The bet seems mostly an excuse for the scriptwriters to assail Martin with bizarre insults ("Your face looks like Robin Williams's knuckles"), but it also reveals a tension within commitment pools. At one point Martin complains, "You can't make fun of me all the time." And his friends respond, "Martin, it's a competition. It's called the Dirty Man Competition. We're going to make fun of you until you shave the beard. That's the rules. That's the whole point." The good news is that the Dirty Man Competition gives Martin an enhanced incentive to keep his commitment: the carrot of being paid, plus the stick of paying if he fails. The bad news is that the competition gives his friends every incentive to see that he fails.

The Dirty Man Competition is a bizarre example of a commitment pool, a kind of pari-mutuel wager where the amount you ultimately pay or are paid turns on how well you do relative to other people in the pool. The NBA's luxury tax—where teams must pay a dollar for every dollar their payroll exceeds the salary cap—is similar. Teams can end up being paid if they are more compliant than the average team, or they can end up paying if they are less compliant than the average team. If all the teams are equally compliant, no money changes hands.

The NBA commitment pool is analogous to anti-charity arrangements, in the sense that forfeited money goes to your adversaries. It would pain the Red Sox (and just about any other reasonable team) to know that part of its luxury tax was going to the Yankees. That's why commitment pools are usually created by friends. After World War II, there were stories of army buddies who put money into a kind of *tontine* agreement to return to France in ten years. Those who didn't show up helped subsidize the drinking of those who did.

We've already encountered Lisa Sanders's $5,000 smoking commitment with her friend. That too was a commitment pool. Lisa can end up being paid if she does better than her friend, or she can end up paying if she does worse. If they both do equally well—either by both smoking or by both not smoking—no money changes hands.

Lisa and her friend could formalize their agreement on stickK simply by each creating a commitment contract that names the other as the beneficiary of any forfeited stakes. But before you enter into any *Biggest Loser*-style competition, think carefully about the Dirty Man. You are giving other people an incentive to make you fail. And all the more, you'd better trust the referees to adjudicate fairly.

A host of other sites are popping up to get a piece of the action. For example, Fatbet.net markets itself as a weight-loss site for competitive people. Fatbet doesn't take your credit card or actually pass along money. But it does make it easy for you and a friend to put your honor at risk. Here's how the site describes its core product:

> All who hit their goals get bragging rights, and those who don't nurse bruised egos until the next round. A Fatbet wager is an additional motivator. Fatbet.net is not a gambling site, so if money changes hands in a Fatbet, we don't really want to know. Here are some Fatbets that *do not* involve cash:
>
> - Losers sing in a Karaoke bar; winners pick the songs
> - Losers let facial or leg hair grow unchecked for 30 days . . .
> - Losers run naked through Seattle Center . . .

We admit that embarrassing Fatbet losers doesn't exactly follow a supportive weight-loss group model, but it can work for the hard core among us who grew up when gym teachers were still mean.

For those who want to take a more cooperative approach to commitment, it is possible to structure group incentives so that individual payoffs turn on the groups' relative success. Robert Jeffrey, the physician who tested the impact of putting $300 at risk if you didn't lose thirty pounds in fifteen weeks, once ran a test to see whether a group contract would do even better than individual weight-loss incentives. In his group contract, the participants were all in it together because refunds were based on the average weight loss of the group. If on average they lost fifteen pounds, they'd each get back half their money. Jeffrey "hypothesized that the group contract would be more effective than the individual contracts" because "the group contract would add a dimension of interpersonal accountability . . . [and] provide a means for individuals not doing well individually to nevertheless contribute to the larger group effort." My gut reaction was just the opposite: that the group commitment would water down the incentive effect because gaining an extra pound wouldn't very strongly affect how much you received. The larger the group, the lower the interpersonal responsibility. That's why when larger groups split the bill at a restaurant, they tend to order more and tip less. In Jeffrey's test, the group that put money at risk had 13 members. I thought this was too large a group to be effective.

But in this small study, both our intuitions were wrong. The group deposits were equal in effectiveness to the individual deposit contracts we discussed earlier. The average weight loss for the group commitment was 31.8 pounds per person—nearly identical to the astounding 32.1-pound success of the individual contracts. And by now, you should realize that a parallel group incentive could be framed in terms of carrots—paying off based on the average success of the group members.

But the bigger lesson of this chapter is that social context matters. Who adjudicates your success matters. Who learns of your success or failure matters. Not all naggers are going to be equally effective. It's not just about you. You're more likely to stick to your guns if you know that people like you have also succeeded. You're more likely to succeed if keeping your commitment is a matter of honor. The big mistake of neoclassical economics was just to ask, "How much money is at stake?" But it also matters who else is affected by your incentive. You're more likely to succeed if your failure punishes the worthy or benefits the unworthy. Anti-charities have no role in traditional economic models, but they can play an important role in keeping you from treating a punishment merely as a price. The behavioral revolution has shown in myriad ways that the question "To whom?" can be just as important as "How much?"

Maintenance and Mindfulness

Ryan Benson's weight-loss story is a cautionary tale for hard-core fans of incentives. In 2005, with the carrot of a $250,000 check dangling in front of him, Ryan shed an amazing 122 pounds, dropping from 330 to 208 pounds in just twelve weeks. That was enough for him to win the inaugural season of NBC's *The Biggest Loser*, and a tearful and much thinner Ryan rode off into the proverbial sunset.

But then the wheels started to come off the bus. Within just five days of the show's end, he had regained 32 pounds. And within two years, he had regained most of the weight he'd lost on the show. When the *Today* show checked in with him in 2008, he was again tipping the scales at over 300 pounds. Many of the *Biggest Loser*'s contestants have parlayed their success on the show into modest careers as motivational speakers and endorsers of weight-loss products. Like Subway's Jared (and, frankly, like me), they have continuing economic incentives to keep the weight off. But Ryan Benson is, if anything, now famous for failing. He's the one who didn't work out.

To my mind, he is the poster child for how not to lose weight. Listen to how Ryan's MySpace page describes his final assault:

> I wanted to win so bad that the last ten days before the final weigh-in I didn't eat one piece of solid food! If you've heard of "The Master Cleanse" that's what I did. It's basically drinking lemonade made with water, fresh squeezed lemon juice, pure maple syrup, and cayenne pepper. The rules of the show said we

couldn't use any weight-loss drugs, well I didn't take any drugs, I just starved myself! Twenty-four hours before the final weigh-in I stopped putting ANYTHING in my body, liquid or solid, then I started using some old high school wrestling tricks. I wore a rubber suit while jogging on the treadmill, and then spent a lot of time in the steam room. In the final 24 hours I probably dropped 10–13 lbs in just pure water weight. By the time of the final weigh-in I was peeing blood.

Few people go to these extremes. But there are seeds of Ryan's mistakes in what many of us do when we try to tie our hands to the mast. We go for the quick fix. With weight loss, we try to lose too much weight, too quickly. And most importantly, we don't have a follow-up plan. This is a recipe for disaster. Ryan went from the high of winning the show to feeling "very depressed because I started falling back into some of the same eating habits I had before the show." He told the *Today* show, "The biggest way it's changed my life is I feel really guilty for gaining the weight back. I know the show inspires a lot of people, so I don't like being the guy to disappoint."

In many ways, Ryan's not to blame. The surreal reality show created great incentives for extreme and extremely short-term weight loss. The rest of us, however, don't have the excuse of "the show made us do it." We are victims of our own goal setting. For the most part, we take goal setting for granted. When people have a problem—say, with smoking or with being overweight—it seems obvious that the goal should be to quit smoking or to lose weight. Economists, especially, jump all too soon to the question of consequences—how big the carrot, how big the stick—and all but ignore the prior question of "Committed to what?"

Andy Mayer is a very different animal than Ryan Benson. "I have a personality type," he said, "that if you give me a goal—whether it be at work or something else—and it is very structured, with key milestones, I can hit it. So, for example, when I decided I wanted to run a marathon, I went out and found out how many miles I was supposed to run. And you know what? I ran exactly that mileage."

Andy's a numbers guy who is constantly trying to figure out how people in the real world will react to changed circumstances. As vice president of consumerology at Express Scripts, Andy spends most of his days trying to figure out how to get people to take their medicine. Ex-

press Scripts is one of the country's largest pharmaceutical benefit managers, helping covered employees be reimbursed for about a half billion
prescriptions a year. I've been working with Andy and some of the
world's best behavioral social scientists on the Center for Cost-Effective
Consumerism advisory board. We're trying to see whether something as
simple as written reminders can help people remember to take their pre-
scribed statin to lower their blood cholesterol. It's an important question
because the problem of therapy adherence is four times greater than the
problem of medical error.

Working with Chicago economist Emily Oster, Andy has been running randomized experiments to find out exactly what kind of reminder
works best. You won't be surprised to learn that reminders pointing out
that "you are less compliant than most other people" are one of the
strategies that help. But what struck me was how much Andy's approach
to his own weight loss differed from his attempts at getting patients to
take their medicine.

At our last board meeting in Washington, D.C., Andy was rightly
pleased with himself for having just lost twenty pounds. And it was especially appropriate that he would want to share his success story with
me, because he had done it using a stickK contract. Early in the morning
on January 1, 2009, Andy, like millions of other people, got up and made
a New Year's resolution to lose weight. But he also knew he needed help.
"Being closely involved in consumerology," he said, "and spending a lot of
time learning about behavioral economics, I knew that if I set a goal with
a commitment bond, that I would hit it or at least make every effort...but I also knew that I needed to tell other people." Andy put
$1,500 at risk and committed to losing a pound a week for twenty weeks.
He also composed an email to send to his friends and family, telling them
about his weight-loss commitment. "I had no trouble signing up for the
contract," he told me. "That was the easy part. The hardest part for me
was telling people . . . That email sat open for several hours. It is one
thing to enter into a contract where you and your spouse know what you
are going to do, but I knew in my head that the social norming part—the
social comparison part—the voyeuristic part was what was gonna make
a difference." Just before dinner, he sent the email, and true to form,
Andy came through with flying colors. I caught up with him five months
later when he was still riding the high that comes with successfully com-

pleting a difficult commitment—in Andy's case, losing almost exactly 10 percent of his body weight.

But Andy was a little taken aback when I pressed him to sign up, then and there, for a maintenance contract. It is hard to lose 10 percent of your body weight. But it's ten times harder to keep it off.

The good news for Andy is that he had taken his time to lose his twenty pounds. A big problem with losing weight quickly is that you don't learn much about how to sustain your goal weight. Andy, by losing the weight more slowly, gained a better idea of what he needed to do in the long term.

The bad news for Andy and millions of others is that taking off weight for three or four months means almost nothing in terms of long-term health. The truth is that many dieters do a pretty good job of losing weight for about half a year, but then their weight tends to drift back toward where it was before the diet. The difficulty of sustaining weight

Figure 11. Weight Loss (and Regain) Under a Weight Watchers Regime

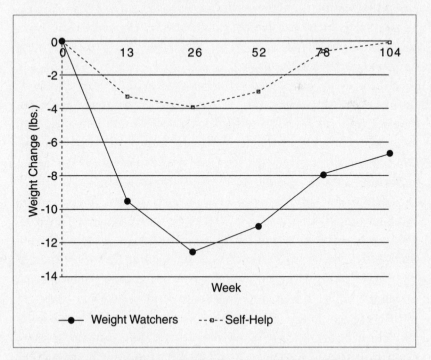

Source: Stanley Heshka et al., "Weight Loss with Self-Help Compared with a Structured Commercial Program: A Randomized Trial," *JAMA* 289 (2003): 1792.

loss can be seen in figure 11, taken from a two-year randomized study of the Weight Watchers program.

The lower, dotted line shows the average weight loss of the Weight Watchers diet group, while the upper, solid line shows the average weight loss for the control group of dieters.

The dotted line should scare the bejesus out of Andy. People on Weight Watchers were able, on average, to lose twelve pounds after six months. But by the end of two years, almost half of the weight was regained. The point here isn't to beat up on Weight Watchers. Study after study of all kinds of diets shows exactly the same pattern: initial success of three to six months followed by a very substantial regain of the weight during the next year or two. The depressing truth is that only about one in five successful dieters—those who, like Andy, were able to lose 10 percent of their body weight—will manage to maintain their weight loss through the end of the year. But if Andy gets lucky and makes it that far, he won't be out of the woods. It takes at least two years to reset your body clock to a new level of consumption—and even after two years, Andy would still have about a 50 percent risk of regaining the weight.

So if you have succeeded in losing a bunch of weight only to put it back on after six months, you are not alone. Among dieters, yo-yoing is the norm, not the exception.

Why do people do so badly after six months? They stop paying attention. We stop getting on the scale for a few weeks and, before you know it, we've gained back several pounds. Then we're scared to get back on the scale. And when we finally do, six weeks later, we've regained half the weight. At that point, we tend to give up. And the cycle begins again. I know. I've not only read the studies, I've experienced this myself.

I begged Andy to enter into a mindfulness contract. The irony was that a person who devotes a great deal of his career to therapy mindfulness resisted embracing mindfulness as a tool to keep the weight off. Even though Andy had just succeeded in using a stickK contract and sincerely believed he was never going to weigh two hundred pounds again, something in him nonetheless balked at the maintenance contract. "I wasn't ready to think about it," he said. "I had gotten to the twenty weeks. I had lost the weight. I needed a little bit of time with the pressure off . . . Do you really want to go out and go for a run a couple of days after a marathon? So, for me it was overkill."

Again, Andy is not alone. At stickK, we offer maintenance contracts that allow successful dieters at the end of weight-loss contracts to promise to maintain their new, healthier weight. But very, very few people sign up for maintenance commitments.

Some of the people who choose to go it alone naïvely believe that they don't need any more commitment help. Flush with confidence over having just lost a bunch of pounds in a few weeks, they think they've turned over a new dietary leaf. If Andy really is going to keep the weight off, it shouldn't be too much of a hassle for him to put some money at risk (since it would never be forfeited). But I worry that Andy's hungry future self is behind some of his resistance, subconsciously stepping in to block anything that might thwart the long-term "feed me" imperative.

Some people are right to resist. It can be downright risky if you are a compulsive eater. A well-known author (and former member of Overeaters Anonymous) confided in me that she was scared to enter into a commitment contract at stickK.com because she was afraid she would lose too much money if she couldn't keep the weight off. But there is much less risk with a promise to report your weight or to regularly attend Weight Watchers meetings. You may not be able to control outcomes, but you can control your inputs.

Since satisfying the dreaded "participation constraint" is so important with regard to maintenance plans, my friend who is scared about putting money at risk on a weight-maintenance bet might do better to reward herself by buying something special if she can maintain her new weight, while simultaneously putting money at risk on some input to healthier living.

What you don't want to do is commit to keeping a diet journal long-term. This may seem like strange advice, because there is good evidence that people who count calories (or points, if you're a WW fan) lose more weight. But the problem is that very, very few people have the mental energy to keep journaling for more than a few months. Counting (and recording) calories can help you lose weight, but it's not a very sustainable strategy for keeping the weight off.

THE SCALE IS YOUR FRIEND

Rena Wing has a better idea. As a professor of psychiatry and human behavior at Brown's Alpert Medical School, Dr. Wing has published more than two hundred articles on obesity—particularly on how to help people like Andy keep from regaining weight. In 1993, she launched the National Weight Control Registry to study people who have been able to maintain at least a thirty-pound weight loss for at least one year. Registrants must initially provide the name of a physician or weight-loss counselor to prove that they actually met the initial requirements. She then checks in with the registrants once a year to see whether they've been able to keep the weight off. Today the registry is following more than five thousand people—that rare breed that has beaten the odds and kept the weight off. (If you are one of these success stories, you, too, can register at www.nwcr.ws.) But Rena follows all registrants year after year, even those who regain all their initial weight. She has painstakingly been creating longitudinal data to find out what behaviors are associated with success and failure.

Some of Wing's findings won't surprise you. Registrants who relapsed were more likely to engage in "emotional eating" and less likely to exercise regularly. But we also know now that those who succeed in the long term are more likely to eat breakfast and watch less than two hours of TV a day. I guess it shouldn't be too surprising that couch potatoes who plonk themselves down on the sofa for several hours each evening are more at risk to pack on the extra pounds. Nevertheless, this is news you can use. It's hard to enter into a contract committing not to eat emotionally. But you could commit to eat breakfast most mornings or to watch less TV.

And most important, Wing has found that people who regularly weigh themselves are more likely to keep the weight off. Regular self-monitoring is the essence of mindfulness. Remember former New York mayor Ed Koch's catchphrase, "How am I doing?" People who establish a routine of asking precisely that are more likely to catch themselves and take corrective action before it is too late. Immediate feedback on when you start to screw up is the key to behavioral modification. The fastest way to house-train a puppy is to instantaneously praise her for pooping outside (and instantaneously chastise her if you catch her in the act inside).

A few years ago, my kids fell into the unattractive habit of speaking "like." "Like"-speak is the young person's dialect where the word "like" is inserted several times per sentence. For example, "I, like, went to the store and tried to buy, like, three Snickers. But the clerk was like 'No way . . .'" I'm not against them speaking "like" to their friends, but I want them to be bilingual. (When supersmart law students are unable to turn "like" off during mock oral arguments, their case invariably suffers.) So to increase my kids' mindfulness, I started to softly say the word "disco" whenever they inappropriately used the word "like." The goal was not to interrupt their speech; it was just to quietly bring to mind their usage. At times I would add a word count by saying "disco one," "disco two" and so on, so that they could find out how many times in just a couple of minutes the word, like a bad weed, was infesting their speech.

If you're having a similar problem with your kids, I highly recommend the "disco" trick. Within a few weeks, the kids were using "like" less often. Sometimes when they misused the word, they would proactively correct themselves by actually saying "like disco." Today, my kids still sometimes speak "like" with their friends, but they know how to turn it off when they are speaking with oldsters.

This kind of mindfulness feedback isn't just for puppies and kids. Maybe the single most useful piece of advice in this entire book is that successful dieters should commit to keep getting on the scale as a crucial first step in trying to maintain their success over time. "If you want to keep lost pounds off, daily weighing is critical," Wing said. "But stepping on the scale isn't enough. You have to use that information to change your behavior, whether that means eating healthier or walking more. Paying attention to weight—and taking quick action if it creeps up— seems to be the secret to success."

This advice flies in the face of those diet gurus who counsel you never to step on a scale. For example, Kelly Minner, one of Ryan Benson's fellow contestants on *The Biggest Loser*, told the *Today* show that as a motivational speaker, she thinks "you should go by how you look and feel, rather than a number on the scale." Some dieters worry that daily weighing will inevitably lead to binge eating and unproductive self-recrimination.

Ultimately, it's an empirical question. It's reasonable to be skeptical of the mere correlation in the registry's longitudinal data. Sure, daily weighers are more likely to be successful. But the causality could go the other way around. It is because you are successful that you are so willing to jump on the scale and record your success. Or it might be some third factor—like persistence—that causes people who weigh themselves to also be successful. As is often true, correlation does not imply causation.

Luckily, Rena Wing has brought randomization to the rescue. In a powerful field experiment, published in 2006 in the *New England Journal of Medicine,* Wing took 314 people who had initial success in losing weight and randomly selected a subset to see if paying attention and taking quick action really caused longer-term success. Her subjects, on average, had lost 42 pounds using Weight Watchers, Atkins, or other similar programs. All of them had lost at least 10 percent of their body weight within the last two years.

She wanted to see if she could strengthen their resolve to keep the weight off. The control group received a quarterly newsletter with tips about diet and exercise. At random, other people were assigned to a "Stop Regain" treatment. They were given a digital scale and told to call in their weight weekly to an automated answering service. They were also urged to plot their progress daily on a chart marked with green, yellow, and red zones. If they were within 3 pounds of their maintenance weight, they were in the green zone and once a month received a small green gift (green tea, a green Frisbee, or a green dollar bill). If they were 3 to 4 pounds over their maintenance weight, they were in the yellow zone. Participants in the yellow zone were urged to tweak their exercising and eating habits to correct a potential problem.

Stronger medicine kicked in if they found themselves in the red zone, with a weight gain of 5 pounds or more: then they received a telephone call from a weight-control counselor and were told to restart their original diet. They were also urged to break open a special red emergency toolbox that contained "their own weight-loss success story, self-monitoring diaries, a book providing information on calories and fat, a pedometer, and several cans of...Slim-Fast." Wing wanted to see whether this green-yellow-red mindfulness program, backed up with proportionate responses, could help keep people on track.

It did. After eighteen months, she paid all the participants to come to a clinic and weigh in. The control-group members on average, had regained about 11 pounds, but the Stop Regain group relapse was only half that—about 5.5 pounds—and its median regain was only 2.5 pounds. Although 72 percent of the control group had regained more than 5 pounds, only 38 percent of the Stop Regain group had ended up in the danger zone.

Ed Messier, from Cumberland, Rhode Island, swears by the program. He initially dropped 46 pounds to get to a very healthy 180 pounds. But he used the Stop Regain program to keep it off. "I use that scale every single day at the same time," Messier told *USA Today*. "That is the most important tool because I'm not going to let myself get 3, 4, 5 pounds out. I won't let myself go."

CHOOSE MY GOAL

I'm rooting for Andy Mayer. But because he doesn't have a firm maintenance plan, I predict that by the time this book is published he will, like so many other "successful" dieters, have gained back at least half of the weight. When I talked to Andy in the summer of 2009 about his future plans, he said he hoped to lose even more weight. I again played Dutch uncle and tried to dissuade him. I told him that empirically it was very hard to sustain even the 10 percent weight loss he had achieved. But he pushed back: "Yeah, but I was more than 10 percent overweight." He wanted to know what people who need to lose more than 10 percent should do. What about them?

It's a good question. But the sad consensus is that without gastric bypass surgery, it is just not possible for most people to lose more than 10 percent of their weight. If you've made bad choices in the past and find yourself weighing 20 percent more than you should, your best feasible option is to take off 10 percent and try to keep it off. As a statistical matter, a large proportion of the excess weight is irredeemably there to stay.

This is a bitter pill for dieters to accept—for the simple reason that most obese people want to lose a lot more than 10 percent of their body weight. A 1997 study of obese dieters found that most wanted to lose more than 30 percent of their initial weight. Most adults want to go back and experience the relative slenderness of their youth. They want to weigh within five pounds of what they weighed when they were twenty (or, if twenty was a bad year, they want to weigh the lowest adult weight they were able to sustain for a year). Unfortunately, for people who have slipped into obesity—those with a body mass index above 30— sustaining a weight loss of this magnitude is very rarely achievable.

Dieters are victims of their own goal setting not just in trying to lose weight quickly and failing to follow through with a maintenance plan.

More basically, we fail to pick reasonable initial goals. Most dieters say they would be dissatisfied with losing even 15 percent of their body weight—and hence are doomed to failure. This is a recipe for disaster.

The health-care community has responded by trying to jawbone dieters into setting more realistic goals. The U.S. Department of Agriculture recommends that dieters try for just a 5 or 10 percent reduction in weight. Health-care researchers stress that there are demonstrable benefits in reducing hypertension, diabetes, and hyperlipidemia from these more modest percentages "even when patients remain considerably overweight."

To bridge the gap between the unrealistic, subjective goals of the overweight and the more realistic goals supported by empirical evidence, some weight-loss counselors suggest a stepped approach. For example, Weight Watchers initially will not let you try to lose more than 10 percent of your starting weight. Only if you lose 10 percent of your body weight are you allowed to set a goal of shedding more pounds. The National Institutes of Health also recommends an initial goal of no more than 10 percent, but adds: "With success, further weight loss can be attempted if indicated through further assessment." Nowhere, however, does it say what would cause further weight loss to be indicated. This reminds me a bit of my childhood goal of learning to play the drums. My parents convinced me that I should learn to play the piano first, and that then I could take on the drums. Come to think of it, I never did have a drum lesson. It would be more honest for the NIH to tell people that they shouldn't attempt further weight loss until they have successfully kept 10 percent of their body weight off for a year (and that very few will make it to their first anniversary). That's what I tried to tell Andy Mayer. But that kind of honesty is just too inconsistent with what people want to hear.

Klaus Wertenbroch, a professor of marketing and behavioral sciences, in 2002 designed an ingenious experiment to test whether students choose the right commitment goals. He placed ads in the student newspaper saying that he was looking to hire "native English speakers to help us proofread papers" and explained that he would pay them 10 cents for every error they caught, less $1 for every day they were late in turning in their assignment. Wertenbroch then used a free Internet text generator to create a postmodern text that was "grammatically correct but not meaningful." He generated tedious essays filled with sentences like: "If one examines semantic discourse, one is faced with a choice:

either accept neocultural desublimation or conclude that the task of the reader is deconstruction, given that consciousness is interchangeable with culture." He then inserted one hundred misspellings and grammatical errors in each of three ten-page essays.

Wertenbroch randomly assigned sixty proofreaders to three different conditions. A third of the proofreaders (the even-deadline condition) were told that one of the essays was due every seven days; another third (the end-deadline condition) were told that all three essays were due in twenty-one days; and the final third (the self-imposed-deadline condition) were told that each of them could choose any deadline for each essay within the twenty-one days.

One of the coolest aspects of Wertenbroch's test is that looking at the choices of the self-imposed-deadline group helps establish whether real people are willing to use commitment devices. Only people who know that they have a willpower problem would be willing to give up some of their flexibility by choosing earlier deadlines and thereby risking higher forfeitures. Many of Wertenbroch's subjects realized they had a problem. The proofreaders showed a clear preference for self-imposed deadlines—voluntarily committing to turn proofreading in before they had to. The lion's share of self-imposed-deadline readers opted for early deadlines, with most opting to space their deadlines out fairly evenly over the three weeks.

I BET YOU CAN EAT JUST ONE

Wertenbroch has done tons of experiments showing that customers are willing to spend extra money to avoid the temptation of overconsuming vices. Regular smokers sometimes buy cigarettes by the pack, even when it would be cheaper and more convenient to buy cartons. In randomized experiments, Wertenbroch has found that consumers are more likely to prefer a larger-quantity discount bag of potato chips to a small bag when the chips are framed as virtue goods ("75% fat free"), but more likely to prefer the small-bag portion when the same chips are framed as vice goods ("25% fat"). Memo to Costco and Sam's Club: stock up on "healthy" versions of the bulk items you are trying to sell. The willingness of consumers to pay more for small-portion packages of vices, like cigarettes, and to pay more for large-portion packages of virtues, like multimonth gym memberships, is a market indication that

at least some consumers know that they have a willpower problem and
are willing to spend money to stiffen their resolve.

What really sets Wertenbroch's test apart, however, is that he can
also tell whether choosing earlier deadlines helped the readers earn
more money. By comparing the results of self-imposed-deadline readers
with the end-deadline readers (who, because of random assignment,
should be equally smart), Wertenbroch was able to ascertain whether
people in the real world choose commitment goals that work, taking into
account the heightened risk of forfeiture. Here the glass is half full. The
self-imposed-deadline group caught more of the errors and was less
tardy than the end-deadline group—and so earned substantially more
money. But the readers who chose their own deadlines did systematically
worse than the readers whose even and early deadlines were imposed externally by Wertenbroch. Of all three groups, the even-deadline proofreaders discovered the most errors, were the least tardy, and (therefore)
earned the most. And all of these differences were statistically significant. Readers who were given the freedom to choose their own commitments were able to make progress on procrastination. But many weren't
able to bring themselves to choose the most helpful deadlines: those that
were evenly spaced. Only readers in the self-imposed-deadline group
who chose evenly spaced deadlines were able to achieve statistically similar results in accuracy, speed, and pay.

Wertenbroch also asked the readers, after the fact, to estimate how
much time they had spent proofreading. And as you might expect, accuracy increased with time spent on the job. The even-deadline group
spent the most time reading (84 minutes per essay), followed by the self-
imposed-deadline group (69.9 minutes per essay) and, finally, the end-
deadline group (50.8 minutes per essay). But the self-imposed-deadline
readers might have been more rational than Wertenbroch gives them
credit for. The even-deadline readers reported liking the task the least,
while the lowest-paid, end-deadline readers reported liking the task the
most. Wertenbroch attributes the difference to the inherent painfulness
of reading postmodern drivel: "These results are not surprising, as the
texts were meaningless and the tasks were boring, if not annoying. We
suggest that the pattern would have been reversed if the task had been
inherently enjoyable." Here, Wertenbroch seems to be forgetting the ten-

dency to dislike obligations that you are paid to do. Another interpretation of the data is simply that the readers disliked the task the more binding they found the deadline. As a group, the self-imposed-deadline readers might have been trading off the prospect of higher pay against the prospect of lower satisfaction when they were considering how early to set the deadlines.

Stepping back, we should see that the glass is at least half full. The self-imposed-deadline readers were victims of their own goal setting in that they earned less than the readers with externally imposed early deadlines. But they made more money than the readers who weren't allowed to set earlier deadlines. All in all, Wertenbroch's ingenious experiment is another case where people choose inefficient goals. But unlike the weight-loss setting, where people err by setting unrealistically aggressive goals, Wertenbroch found that his readers erred by not being willing to set sufficiently aggressive goals.

The threat of losing money—our old friend loss aversion—may have chilled some of the proofreaders' goal-setting ardor. But this chilling effect, when applied to weight loss, might be a good thing. The pie-in-the-sky weight-loss resolutions to once again fit into your college clothes are usually made without any threat of monetary loss. Instead of trying to jawbone dieters into setting more realistic goals, it might be helpful to have them put their money where their mouth is. Big talk is easier when it's cheap talk.

I thought very differently about my goal weight when I decided back in 2007 to risk $500 a week. I wanted to ultimately get to 180 pounds, but I set my forfeiture weight at 185 so that I would have some wiggle room. I wanted a warning track of five pounds to let me take corrective action before I hit the forfeiture wall. Turns out my intuition has some scientific support. Five pounds is about the amount that an adult's weight normally fluctuates—taking into account things like meals, menstruation, and exercise. That's why Rena Wing's Stop Regain program sounded an alarm when a maintainer's weight exceeded this amount.

ILLUSORY PROGRESS AS AN INCENTIVE

Wing's five-pound buffer is such a good idea that you might even try using it at the beginning of your diet. You might wear particularly heavy clothes (and even put a couple of rocks in your pockets) before getting on the scale at the beginning of your next diet. If you've put money at risk, like Andy did, to lose a pound a week, the heavy-clothes trick will build some wiggle room into your very first weigh-in. That way you almost certainly will experience initial success when you get on the scale the next time. Just creating the illusion of progress can help speed people toward their ultimate goal.

Columbia Business School professor Ran Kivetz used a local coffee shop to prove that this is true. He convinced Café Cappuccino to run a field experiment on the impact of using two different reward cards (see figure 12).

Figure 12. Coffee Cards That Are More Different Than They First Appear

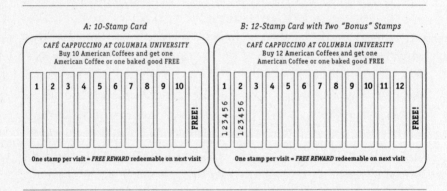

The ten-stamp card asked patrons to accrue ten stamps before they got a free coffee (or baked good), while the other card required twelve stamps to get a free item, but came with two of the twelve slots already stamped. While the two cards imposed substantively identical incentives, the twelve-stamp card gave the illusion of progress because these card holders started out one-sixth of the way to their goal.

Kivetz's inspiration for the experiment came from rats. In the early 1930s, Yale psychologist Clark Hull noticed that rats on a straightaway "run faster as they near the food box than at the beginning of the path." Kivetz had found a similar effect when analyzing nearly ten thousand purchases by customers using the café's original ten-stamp card. Patrons averaged 3.5 days between their second and third stamps, but only 2.5 days between the last few stamps.

Kivetz wanted to see if merely giving people the illusion of progress would be enough to get them to behave like Hull's rats and speed up as they seemed closer to the reward. It's pretty insulting to think that Columbia students would be tricked by smoke and mirrors. But the rat race worked. Randomly selected patrons who received the twelve-stamp card reached their goal of getting a free coffee nearly 20% quicker. Customers who completed the ten-stamp card averaged 15.6 days, while customers who completed the twelve-stamp card averaged only 12.7 days.

Of course, it might be harder to trick yourself with progress illusion. If you wear a twenty-pound backpack to establish your before-diet weight, then take it off for your first weigh-in, you may have trouble convincing yourself that you are already 50 percent of the way toward your goal. Then again, it probably couldn't hurt.

Behavioralists even have something to say about when you should emphasize the accomplished or unaccomplished portion of the goal. Do the Columbia students drink more frequently because they know they are 75 percent of the way there? Or is it because they know they have only 25 percent left to go? This is the kind of question the neoclassical economist would never ask. One bit of information implies the other; to a rationalist, they are different sides of the same accomplishment coin.

But to a behavioral psychologist like Ayelet Fishbach, this question is ripe for investigation. Fishbach sent out hundreds of letters soliciting contributions to Compassion Korea, a Christian child-sponsorship organization that was in the midst of a campaign to raise 10 million won (approximately $10,000) "to help AIDS orphans in Africa." Some of her

Figure 13. Framing Contributions as "To Date" Versus "To Go"

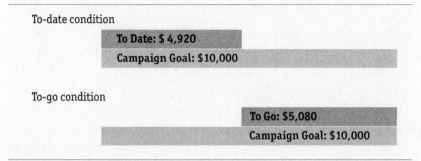

To-date condition

To Date: $ 4,920
Campaign Goal: $10,000

To-go condition

To Go: $5,080
Campaign Goal: $10,000

Source: Minjung Koo and Ayelet Fishbach, "Dynamics of Self-Regulation: How (Un)accomplished Goal Actions Affect Motivation," *Journal of Personality and Social Psychology* 94 (2008): 183.

letters included graphics emphasizing how much had been raised to date, while other letters emphasized how much money was still left to be raised. Figure 13 shows the two key graphics, translated from the Korean. Which type of letter do you think provided the stronger incentive?

Your response might say a lot about how committed you are to helping AIDS orphans in Africa. Ayelet found that people who had previously given to the charity were much more likely to respond to the "to go" graphic. Only 1.6 percent of the existing contributors who received the "to date" letter made a pledge, whereas 12.5 percent of the existing contributors who received the "to go" letter made an additional pledge. However, the incentive flipped with regard to recipients who had expressed interest in the charity but had not yet given before. These prospective donors were much more likely to respond to the "to date" letters than to the "to go" letters (24.2 percent versus 8.3 percent). Overall, emphasizing the right message increased the expected contribution more than threefold.

Ayelet thinks that focusing on the unaccomplished portion of a goal provides a bigger incentive for those who have a deeper commitment to the goal, while focusing on the accomplished portion better motivates people whose commitment is more uncertain. She's backed up her intuition with her own study of loyalty reward cards. Students at the University of Chicago are more likely to express interest in a bookstore rewards card concerning products they are committed to buying (textbooks) if the rewards card emphasizes the "to go" purchases, but they're more interested in reward cards for discretionary items (coffee mugs and other university paraphernalia) if the rewards card emphasizes "to date" purchases.

Ayelet's research has real implications on how to best motivate yourself when you're in the middle of trying to reach a goal. For example, imagine that you're trying to pay off your credit card bills. If you are less certain about how committed you are to the goal of being debt-free, you're probably better off emphasizing your progress to date—and maybe even fostering an illusion of progress. But if you're deeply committed to the goal of getting out of debt (possibly because you've put a lot of money or honor at risk), you're probably better off emphasizing how much you still need to pay. For the truly committed, it is better to focus on how many miles you have to go before you sleep.

A FLEXIBLE COMMITMENT IS NOT AN OXYMORON

Mortgage contracts are a classic commitment device: make your monthly payment or get kicked out of your house. At least, that's how they are supposed to work. But in Australia, New Zealand, and the United Kingdom, millions of home owners have opted for a new kind of "flexible mortgage" that gives them the option of skipping a payment from time to time. If you've just spent a bunch of money on Christmas presents or on a wedding for one of your kids, you can take a payment holiday for a month or two. Your mortgage accrues unpaid interest during the hiatus, but you don't go into default. As one happy customer explained, "I [asked] for a mortgage payment holiday to help pay for our actual holiday. We had an absolutely brilliant time with no bills to worry about when we came home. I'd never go back to a traditional mortgage now."

The flexible mortgages can tell us a lot about choosing the right commitment. Some people have the mistaken idea that flexibility is antithetical to the whole idea of commitment. But in setting the right goals, it is important to think in advance about building in the right amount of flexibility. A five-pound warning track is one way to give yourself some leeway, but there are tons of others. As with the flexible mortgage, you might just exempt yourself from compliance a certain number of times per year. For some commitments, flexibility is probably not a good idea. Having quit smoking, Lisa Sanders is better off committing not to have another cigarette, no matter what. But people who are still in the process of quitting might reasonably allow a few upticks in their consumption— so long as the overall trend is downward. And our New Zealand friend James Hurman exempted cigars in part because he wanted to be able to celebrate the birth of his children.

Chapter 3 spoke about reward and penalty schedules that contemplated the possibility of failure. A standard incentive for heroin abstinence uses increasing rewards for prolonged clean tests, but resets the reward back to the initial, lower level if a dirty test is submitted. Some penalty schemes start small, to give the offender a taste of some unpleasant consequence, and escalate only if there are repeated instances of noncompliance. But an alternative approach is to define the goal so that episodic deviations do not trigger a penalty (or a reduction in the prospective reward).

People who do go forward with long-term commitment contracts often have trouble thinking ahead and experience commitment fatigue. If you commit to exercising three times a week for the next sixth months, are you going to get in trouble if there is a big project at work or school that gets in the way? Sometimes just verifying and reporting your success can become a hassle.

The downside of flexibility is that it can be abused. An alcoholic who makes an exception for a special occasion can all too easily fall off the wagon and spiral downward. And for some one-shot commitments—like writing a will—it is best to choose an inflexible commitment but set the date sufficiently far in the future so that you know that you have time, even taking into account the vagaries of job and family demands. In fact, if you are reading this and don't have something basic like a will, put down this book and create a commitment contract to go have a will made in the next six months (and back up your commitment by putting some money at risk on your credit card and naming a loved one—or even your lawyer—as a referee). It will take five minutes now to create the contract and will radically increase the chance that you'll end up with a will.

Many people resist "wasting" the five minutes. They figure they'll just find the time to contact a lawyer and have a will made. But if it's taken you this long, and you still haven't acted, why do you think that the next six months is going to be any different?

In engineering the optimal amount of flexibility into your commitment, it's vital not to throw the baby out with the bathwater. Your exceptions should themselves be hard-edged. The law makes a distinction between "rules" and "standards." Rules are specific commands: Thou shalt not drive more than forty miles per hour. Standards are more amorphous guides: Thou shalt not drive recklessly. With commitments, there is a wealth of evidence suggesting that more specific goals are more likely to be more effective. So you shouldn't enter into a commitment contract to "make progress" on having a will made. You should commit to actually having a will made, or to at least calling a lawyer by a certain date.

The big problem with standards is that they require a nontrivial assessment after the fact of whether or not you succeeded. This puts a lot of pressure on the referee. If you referee your own contract, it's often going to be too tempting for your future self to tell a story as to why you satisfied the standard. After all, it's your future self who is trying to put off doing the unpleasant task. If you choose someone else to be your referee, committing to a standard is a recipe for conflict and bad feelings. Dean Karlan has the dream of harnessing the wisdom of crowds on the Internet as an objective adjudicator of whether a standard has been met. Thus, I might commit to being able to play a "reasonable rendition" of "Für Elise" by the Fourth of July and then let a YouTube clip of my performance be submitted to a candid world.

But some things aren't susceptible to YouTube adjudication or specific commitments. I've successfully used commitments to help me do all kinds of things—lose weight, exercise, read novels, even write this book—but these aren't my only behavioral problems. I'd also like to be more caring toward my spouse. Sure, I could commit not to drink out of the milk bottle, but the more important improvements I'd like to make concern the ways I speak with her.

Even more, commitment contracts can't change the way you feel. A commitment not to think about pink elephants is virtually guaranteed to fail. But even with affairs of the heart, you can help the way you act. Shortly after stickK went live, a user custom-designed a Do Not Call commitment that deeply resonated with some of my own years in the dating wilderness. She promised not to call Henry "unless he asks me to, leaves a message, or I missed his call." She titled her commitment "Don't make someone a priority if they only make you an option." This commitment (which she backed up by putting a small amount of money at risk) was an important step in helping her move on with her life.

EVENTUALLY

In the spring of 1943, the metaphorical progenitor of all behavioralists, B. F. Skinner, was toiling away not in a university lab but on the top floor of the Gold Medal flour mill in Minneapolis. Skinner's lab looked out on a three-story neon sign that simply said, "Eventually," referring to Gold Medal's marketing slogan, "Eventually, you will use Gold Medal flour." Skinner, to aid the war effort, had taken a leave from the University of Minnesota to set up a secret lab funded by General Mills to see if he could train pigeons to guide missiles. The research was classified as "Project Pigeon."

Skinner's biggest breakthrough came on a day when he was waiting around for Washington to sign off on the next stage of his research:

All day long, around the mill, wheeled great flocks of pigeons. They were easily snared on the window sills and proved to be an irresistible supply of experimental subjects . . . One day we decided to teach a pigeon to bowl. The pigeon was to send a wooden ball down a miniature alley toward a set of toy pins by swiping the ball with a sharp sideward movement of the beak. To condition the response, we put the ball on the floor of an experimental box and prepared to operate the food-magazine as soon as the first swipe occurred. But nothing happened. Though we had all the time in the world, we grew tired of waiting. We decided to reinforce any response which had the slightest resemblance to a swipe—perhaps, at first, merely the behavior of looking at the ball—and then to select responses which more closely approximated the final form. The result amazed us. In a few minutes, the ball was caroming off the walls of the box as if the pigeon had been a champion squash player.

Originally, Skinner thought that the best way to train a desired behavior was to wait until the animal eventually did it of its own accord, and then to immediately reinforce the behavior by giving the animal food. The problem with this approach is that more complicated behaviors, such as pigeon bowling, almost never spontaneously occur. You could put monkeys in front of typewriters, but you'd have to watch for an awful long time before one of them decided to type *Hamlet*. Skinner's Gold Medal breakthrough was discovering a new meaning of "eventually." Instead of waiting for a multipart behavior to occur "eventually," on the animal's schedule, Skinner could train an animal much more quickly by waiting only until the animal literally nodded its head in the right direction. He trained the pigeon through a series of small events—by reinforcing behaviors that were "successive approximations" of the desired behavior. Using this new technique, which he called "shaping," Skinner had rats patriotically sitting up on their hind legs, hoisting the Stars and Stripes, and saluting until the national anthem finished playing.

Once again, we may be able to learn something from rats and pigeons. Breaking up our goals into smaller subcomponents not only can give us a more realistic chance of success; it can condition and shape our preferences—teaching us new patterns for operating. But we also know that it will work only if we can will ourselves to follow through on the subsequent parts of the training. Skinner's pigeons and rats didn't have to worry about follow-through. Skinner was there, doling out the reinforcements time and time again. For humans to succeed at profound change, we have to be willing to go beyond the quick fix. Committing to longer-term mindfulness is a powerful strategy to aid in this endeavor.

Of course, choosing an effective goal will work in the future only if we can bring ourselves to commit to it in advance. But as we've seen, the participation constraint often binds. We often fear committing to the kinds of substantive goals that would later on be the most effective. Part of this fear is unreasonable; it's your future self discouraging your present self from committing to real change. But part of the fear is eminently reasonable. You just can't know whether something will come up to get in your way. You might plan to lose weight, but then become pregnant. You might plan to save money, but then lose your job. Choosing the right goal, with the right degree of difficulty as well as flexibility, is crucial in garnering your initial consent to be governed by the device in the first place.

What Commitments Say About You

People who say they're going to be brief often aren't. We've all been there. In a meeting, someone will announce that she's going to brief, only to drone on and on. When I hear a speaker say, "I'll be brief," I'll frequently time the oration, as well as the next speaker's. I find, as often as not, that the self-proclaimed "brief" speaker holds the floor longer.

The three syllables are in one sense self-contradicting. By taking the time to utter them, you are not being as brief as you might have been. Wouldn't it be better to forgo this temporal throat clearing entirely? But my peevishness about this, which is at most a venial sin, seems at odds with the larger themes of this book. "I'll be brief" is a kind of commitment. Speakers uttering these words are giving the audience permission to cast aspersions toward them if they speak for too long.

The problem is that "I'm going to be brief" is just cheap talk. The consequences of speaking too long are so wimpy that they don't seem to constrain. This is a place where a more specific commitment is needed— something that has a number attached to it.

One problem is that people underestimate how long they speak. This is a concern at Quaker meetings, where individual congregants largely self-regulate how long they are going to talk. If you asked speakers immediately upon finishing an oration to estimate how long they spoke, I bet that many people would report that they spoke for fewer minutes than they actually did.

One solution to this problem may be for groups to develop a norm to have speakers publicly estimate how many minutes they intend to speak before they begin. Instead of saying, "I'm going to be brief," it would be a stronger commitment to say to the moderator, "Please tell me if I speak for more than X minutes."

This is a commitment that has some teeth to it. It's more credible. A friendly audience is unlikely to interrupt you if you exceed your self-imposed limit by a minute or two. But the likelihood that an audience member will interrupt you increases as you continue to monopolize the airtime. Just the threat of being called out will keep some people in line, and the worst offenders will learn that they're speaking for longer than they thought.

Speakers droning on at conferences and meetings isn't one of the biggest problems in the world. But this simple verbal commitment is different from many of the earlier examples in this book because it is a commitment that is directed much more toward others. When I commit to reading twenty pages of a novel every day, I am doing it mostly for myself. To be sure, many personal commitments have mixed motives. Lisa Sanders wants to quit smoking both for herself and for her kids. I want to lose weight in part to look more attractive for my spouse.

But this chapter's focus is on commitments that are more dominantly directed toward others. Commitments signal to others that you really mean to do something. We already saw an example of this in my failed attempt to get Simon Usborne, the reporter from the *Independent*, to follow through and call me for a book interview. But the idea of appointment commitments has much broader applicability. I asked my dentist whether he ever had the Usborne problem—if he'd had patients skip one or two appointments and still ask for another. He grew suddenly animated and answered, "One or two? We have patients who miss five or six appointments in a row, and still want us to schedule another. At some point, we just say no." But instead of ending the relationship, a more moderate solution would be to demand a forfeiture if they miss again. Some doctors' offices already do this. (In the movie *Annie Hall*, Woody Allen's character jokes, "I was suicidal, as a matter of fact, and would have killed myself, but I was in analysis with a strict Freudian, and if you kill yourself, they make you pay for the sessions you miss.") And the beautiful thing about appointment commitments is that they can be mutual. I might have had more success with Usborne if I had instead suggested that we *both* put £50 at risk if either of us failed to be available at the appointed time. I'd feel better about paying my doctor for being late if he did me the same courtesy.

When the British airline easyJet promises, "Our fares will be the lowest available on any route [or] we'll refund you double the difference!" it is signaling to harried customers that they don't need to spend time searching for the lowest fare. The essence of contracting is the idea of making binding commitments. But normally a contract is just a promise to perform or to compensate if you fail to perform. In the language of chapter 2, expectation damages are merely incentives. Promisors can later decide whether it is cheaper to compensate than to perform. In contrast, "double the difference" promises do more than merely compensate—they're a signal that easyJet really intends to pay attention and make sure its prices are the lowest. In March 2008, easyJet even promised to refund triple the difference if it was undersold. But beware of the fine print. Britain's Advertising Standards Authority ruled that the ads were deceptive and enjoined their future use—in part because one contractual hand took away what the other gave. The fine print limited refunds to £100 and required all claims to be made within an hour of booking.

BEHIND THE COMMITMENT SCREEN

In the early 1990s, my hometown of Kansas City was being ravaged by crack. Like falling dominoes, a line of poor neighborhoods were being decimated one after another by crack houses and crime. I was asked if I could help by buying a duplex that would be dedicated to low-income tenants to help shore up one of the at-risk neighborhoods. The idea was to scatter twenty brand-new duplexes in the neighborhood and create a buffer of sorts. The motivating force behind the project was Richard Eisner, who was both the duplex builder and the units' manager. I remember how impressed I was with Eisner's behavioral smarts. The duplexes were really nice, with central air-conditioning and lots of surprising details. But, oddly enough, they didn't come with stoves. Eisner explained that a used stove cost only about fifty bucks, but that tenants who went to the trouble of installing one were much more likely to stay around. Eisner had never read Thaler, but he understood that losses loom large. This was a guy I wanted for my manager.

Under the state program, I was committed to owning and renting the duplex for fifteen years. My biggest concern was that Eisner might move on to greener pastures. So when I was negotiating the purchase price, I tried something different. I asked Eisner to speculate on how much he thought the duplexes would be worth in fifteen years, at the end of my commitment. Eisner reasonably said he thought they'd be worth $100,000 each. So I asked him if he'd be willing to give me a put option to sell my duplex back to him for $50,000. Eisner would be taking the risk that they would end up worth less than $50,000. But based on his own estimate of the duplex's value, it wouldn't be much of a risk, and crucially, as manager, he would have a lot more ability to affect its value than I would have as an absentee owner. Eisner refused my offer and wouldn't even consider giving me an option to sell the duplex back for $20,000.

I should have learned something from his refusal. I went ahead and bought the duplex and, sure enough, a few years later Eisner walked away from managing them. Commitments can be valuable for what they say to others about us. But they're also useful as screening devices. Beware of the person who swears that something is going to happen but isn't willing to put his money where his mouth is. Consider the repentant husband who swears up and down that he will stop his philandering. If he refuses to sign a postnup forking over more of the community property if he strays again, you should worry that he isn't really serious about changing his low-down cheating ways. Think of it as a sincerity filter.

Years ago I helped represent a convict who was on death row. The convict's main lawyer (who, because of attorney-client privilege, must remain anonymous) was working for free because he strongly opposed the death penalty. However, in taking the case, the lawyer was bound to not only help overturn the death sentence; he also had to try to overturn the underlying conviction and win his client's physical freedom. The lawyer was worried that if he succeeded in getting his client back on the street, the guy might kill someone. That's where I came in—and the wonders of commitments as screening devices. I drafted a contingent-fee agreement stating that the legal representation in the case would be free only so long as the convict stayed on the straight and narrow. If the convict was ever charged with a violent crime or even possession of a firearm, the full fees would immediately come due. What's more, the agreement also included the convict's parents and expressly put their house at risk to pay the fee. This contingent-fee agreement (which the convict and his parents immediately agreed to) not only gave the convict and his family an extra incentive to avoid the commission of violent crimes but morally protected the lawyer. If the unthinkable happened, the lawyer would at least know that he had not given away his services to help someone return to a life of crime.

When people are trying to convince you that they have changed, it is often very hard for them to resist making commitments—particularly if you let them choose the fallback terms. For example, when I was buying customized Formica modular tabletops for my home office, I made sure to ask when the furniture store thought the tabletops would be installed. The salesman said within two weeks. That sounded great to me, but I explained that I'd been burned by a couple of late furniture deliveries and wanted him to specify a date after which I could cancel the contract. He responded by saying that if they were not installed in my house within two months, he'd sell them to me at half price. And crucially, we jotted this new condition down right on the sales order. I didn't hear from the company again until five days before the two-month deadline. The store called pleading for a few extra days. I politely declined, and the tabletops were installed with one day to spare.

Beyond tabletops and dentist appointments, commitment screens could also help the government assess all kinds of things. In 2007, when Miller and Coors proposed to join forces, they said to antitrust authorities what merging parties always say: The merger will reduce our costs and make us a lean, mean, competitive machine. But antitrust authorities had to worry that the merger would make it all too easy to collude in a world where just two firms—Anheuser-Busch and MillerCoors—would control roughly 80 percent of the U.S. beer market. What's a regulator to do? One possibility was to ask MillerCoors to sell competition bonds that would force the newly merged company to pay out if the merger ended up being anti-competitive. The key to figuring out whether the merger was ultimately anti-competitive would be to look at what happened to Anheuser-Busch. If the price of Bud dropped after the merger, the merger was probably a good thing. But if the inflation-adjusted price of Bud started going up, MillerCoors would have some explaining to do. Or the Justice Department might have approved the merger only if MillerCoors was willing to sell call options on Anheuser-Busch stock. If the stock price of Anheuser-Busch rose more than the S&P, MillerCoors would have to pay out a penalty to greedy option holders. Selling (at the money, index-adjusted) calls on its competitor gives MillerCoors less of an incentive to collude on the price of beer. Any money it made from selling its own beer at a higher price would be at least partly offset by what it would lose in paying off on the competition bond. What's more, the Justice Department, by demanding competition bonds, can help screen out the worst merger attempts. Regulators should be skeptical about the sincerity of company claims that a merger will be pro-competitive if the company is not willing to stand behind the promise by paying a penalty if the anti-competitive results ensue.

IT'S YOUR JOB TO BE HEALTHY

Doug Short has the graying hair of a man in his early fifties but the face and athlete's body of someone half his age. Thirty years ago, when he was in college, he ruptured a disk playing basketball that changed his life. After three operations, he found that if he did not exercise almost daily and warm up his joints, his back would get seriously stiff during the middle of the day. Regular exercise has for years been a part of his unvarying routine. "In my case," he told me, "it is really because of that thorn in the flesh that I changed." He knows from personal experience that changing your lifestyle can have a profound impact on happiness. And as the president and founder of the BeniComp Group, he put this knowledge into action.

BeniComp Advantage has caused a stir in the group insurance industry by providing companies with plans that lower an employee's deductible by hundreds of dollars each year if the employee meets certain National Institutes of Health wellness goals. "It's like a good-driver discount for health insurance," Short said. For example, in Benton County, Arkansas, the BeniComp plan came in and raised health-care deductibles for county employees from $750 in 2004 to a whopping $2,750 in 2005. But the employees can reduce their deductibles to as low as $500 if they don't smoke, are not overweight (with a BMI of less than 24.9), and have low cholesterol (LDL under 160), low blood pressure (lower than 140/90), and low glucose (under 126).

The potential expense of $2,200 a year is a powerful carrot to change behavior, but a common initial reaction is that an employee's health is none of the employer's business. Andy Bowman, a juvenile detention officer for Benton County, said, "I didn't like it. I didn't want no one telling me I'm that bad out of shape." Lewis Maltby, president of the National Workrights Institute, views the incentive as a form of lifestyle discrimination. "You are supposed to be paid on the basis of how you do your job, not how often you go to the gym or how many cheeseburgers you eat," Maltby said.

But getting employees to be healthy is the employer's business because health-care costs have become such a dramatic part of the costs of employment. In 2009, the average cost for a company to cover a family was $12,680. President Obama is now emphasizing that this is nearly the same as the full-time minimum wage. More importantly, the lifestyle choices of employees outside of work actuarially drive how much employers will have to pay in health-care premiums. As much as 70 percent of health-care costs are related to lifestyle. Employees who smoke or who are overweight cost their employers $2,000 to $3,000 a year *extra* in health care and sick days. When you eat that extra cheeseburger at night, you are imposing real costs not just on your employer but on your fellow employees. With employer health-care costs spiraling out of control, employers have reacted by reducing coverage and increasing co-pays and deductibles for all employees. Workers who make unhealthy lifestyle choices aren't just hurting themselves.

Increased co-pays and deductibles incentivize workers to stop going to the doctor. But BeniComp incentivizes them to see the doctor—first at annual screenings to check their core health metrics and also later on. Employees whose cholesterol is too high have a powerful incentive to get a prescription for statins. Employees whose blood pressure is too high have an incentive to take ACE inhibitors or beta blockers. "What we've discovered is when I put lifestyle factors in on the health plan, the dental spend[ing] went down," Short told me. "And . . . the single biggest lifestyle [factor driving] dental claims is gum disease which was driven by smoking." Short also said that when body mass was included in BeniComp's assessments of employees' health, "we discovered that the vision [spending] then started dropping. Because the body mass was the leading . . . [cause] of diabetes."

In 2005, Benton County's health-care fund was $480,000 in arrears. But after just one year, the fund was showing a surplus of more than $1 million. "More people are being treated, but it costs half as much per claim," Thomas Dunlap, the county's benefits administrator, said.

Benton County is at the cutting edge of a major movement toward employee health-care incentives. By 2007, nearly half of major employers were using incentives to induce employees to participate in wellness initiatives. Often these incentives are small gifts distributed for answering online health-screening questionnaires. But other companies are cracking down on unhealthy lifestyles. In 2007, Black & Decker started charging smokers $25 more a month for health care. Whirlpool is even more severe—charging an extra $500 a year for employees who smoke. In 2008, it suspended thirty-nine workers for lying on their benefits enrollment forms to avoid the surcharge.

Others employers, like the Cleveland Clinic, are refusing to hire smokers. Scotts Miracle-Gro has simply stopped employing smokers altogether and backs it up with random urine tests for nicotine. Taking away your job is the ultimate stick an employer can wield to induce change.

Whether the incentive is framed as a carrot or a stick can make a big difference. The Tribune Company caused an outpouring of employee protest in January 2008 when it started requiring employees to pay $100 a month more in insurance premiums "if they or any of their covered family members smoke." Just four months later, Tribune yielded to pressure, rescinding the policy and refunding past surcharges. BeniComp insists on framing its approach as a carrot (although it is a reward only once the employer moves to a huge deductible). "You don't want to be punitive," Short said. Jim Winkler, a national health-care-practices leader at Hewitt Associates, agrees. "If you market an incentive or a discount as a positive for people, it is much more broadly accepted," he said. The one drawback of the carrot frame is that it is harder to reverse. "If [Tribune] had put the same mechanism in place but marketed it as a discount for nonsmokers," Winkler said, "they would have had a firestorm of people objecting about having it taken away."

Regardless of whether the incentive is characterized as a carrot or a stick, is it fair to charge different amounts for things that are beyond an employee's control? The Department of Labor doesn't think so. In late 2006, it responded to the rise of incentivized employee wellness programs by adopting new HIPAA (Health Insurance Portability and Accountability Act) regulations limiting the kinds of incentives that can be offered. To be compliant, an employer's plan must be adequately explained and provide at least annual opportunities for employees to qualify. Additionally, employers must give employees a reasonable alternative way of qualifying for a carrot (or avoiding a stick) when achieving a goal is "unreasonably difficult . . . due to a medical condition or medically inadvisable." It's obviously inadvisable to try to get pregnant women to lose weight. The tougher question is trying to determine when a medical condition makes it "unreasonably difficult" to achieve a particular goal. But BeniComp has had few problems in offering reasonable alternatives, because very, very few employees come forward with sufficient evidence of difficulty. Short explains, "If somebody called in and said . . . 'I just can't do it,' my first response to that appeal is 'I need a letter from your doctor that says it is medically inadvisable for you to change.' Rarely do they come back."

The HIPAA regulations also cap the total value of rewards that can be made using a figure contingent on health-care standards. A wellness incentive can't be more than 20 percent of the total cost of health-care premiums. I understand the need for a cap, but it's hard to know whether 20 percent is the right one. Safeway, for example, estimates that a family with two overweight smokers under its health plan will cost it an additional $4,480 to insure, but under the cap Safeway is allowed to offer annual incentives only up to $1,560 (20 percent of $7,800, the yearly premium for this family). The law doesn't allow Safeway to offer a $4,000 bonus in order to save $4,480. To make matters worse, the more successful a wellness incentive is, the less the employer can offer in the future. If Safeway offers a $1,500 health bonus and its annual premiums decrease to $5,000, it will have to reduce its offer to a paltry $1,000. Safeway's CEO, Steven A. Burd, wants change he can believe in. "Reform legislation," he said, "needs to raise the federal legal limits so that incentives can better match the true incremental benefit of not engaging in these unhealthy behaviors."

Help is on the way. Congress is likely to give employers incentives for wellness incentives. It is less clear whether the 20 percent cap will be lifted. But several members of the Obama administration really get the power of behavioral economics. Peter Orszag, Obama's budget director, has long promoted the value of using incentives to encourage healthy lifestyles. And he puts his money where his mouth is: Orszag used stickK.com to help him train for a marathon. "If I didn't achieve what I wanted to, a very large contribution would automatically come out of my credit card and go to a charity that I very much didn't support," Orszag told *The New Yorker*'s Ryan Lizza. "So that was a very strong motivation, as I was running through mile 15 or 16 or whatever it was, to remind myself that I really didn't want to give the satisfaction to that charity for the contribution." Orszag has been diplomatically silent about which anti-charity kept him running. But Lizza thinks "it's safe to say that if Orszag doesn't do well in his next marathon, the Bush library can look forward to a nice donation."

DOING WELL BY COMMITTING TO DO GOOD

Employee commitments provide ways that workers can speak to their bosses. But commitment speech can run both ways. Companies regularly use contractual commitments to speak credibly to their employees and customers. Honest Tea promises that it's Peach Oo-la-long is a Fair Trade–certified product. You can feel proud about working at the Gap, because your employer has been accredited as an SA8000 (socially accountable) corporation.

Jennifer Gerarda Brown and I have tried to use the same commitment-certification idea to promote equal employment rights for gay workers. (Jennifer is a law professor at Quinnipiac University, and we entered into the granddaddy of all commitments back in 1993 when we married.) In 2005, we noticed a huge disjunction between public sentiment about equal employment rights and federal policy. Gallup polls showed that a whopping 88 percent of Americans felt that "homosexuals should . . . have equal rights in terms of job opportunities." Americans may still be divided on gay marriage, but there is an overwhelming consensus against employment discrimination. Yet in most states it's still legal to refuse to hire someone because he or she is gay. At the federal level a modest bill, the Employment Non-Discrimination Act (ENDA), has been introduced several times in Congress without success. ENDA is less far-reaching than other federal civil rights statutes—it prohibits only conscious employment discrimination on the basis of sexual orientation. A boatload of prominent corporations—including the likes of AT&T, MillerCoors, IBM, and General Mills—have already endorsed ENDA. But Congress, to date, has been unwilling to give them what they want.

That's where Jennifer and I come in. Because of us, firms that are committed to employment equality for gay and lesbian workers don't need to wait for state or federal legislation. We created a certification mark that allows individual corporations to privately commit to nondiscrimination. You already know from chapter 4 that I've helped Equality Forum nudge Fortune 1000 firms to include sexual orientation in their nondiscrimination policies. But the pretty words of nondiscrimination policies sometimes turn out to be only that. If an employer discriminates against an applicant because she is gay, it is far from certain that the employer would be liable for breach of contract—even if the employer has a nondiscrimination policy. The nondiscrimination policy is the civil rights analogue to a speaker saying, "I'll be brief." It makes you feel good when you say it, but it's too vague to have much bite.

So instead of merely promulgating nondiscrimination *policies,* Jennifer and I tried to figure out a way for corporations to adopt nondiscrimination *promises,* binding contractual commitments that replicate the rights and responsibilities of the civil rights legislation that has yet to become law. The United States Patent and Trademark Office granted us the right to license a new certification mark, which is just the letters FE inside a circle:

Figure 14. The Fair Employment Mark

We call it the Fair Employment mark, and by licensing it, employers commit to the exact substantive duties of ENDA. With just a few clicks of the mouse at www.fairemploymentmark.org, any employer can turn its policy into a legally enforceable promise.

The idea is simple, really. By signing the royalty-free license agreement, an employer gains the right (but not the obligation) to use the mark, and in return promises to abide by the word-for-word strictures of ENDA. The license gives employees the right to sue for violations of the act, seeking the core compensatory remedies that ENDA would, if enacted, give them. (We did not try to improve on ENDA here, just to copy it. Since

ENDA requires a filing of claims within 180 days, so does our license. Since ENDA would allow arbitration agreements, our license would as well.)

Why would an employer voluntarily take on greater vulnerability to lawsuits? First and foremost, committing to equality is good business. Displaying the mark on a product or an advertisement of service signals to knowing employees and consumers that the company has committed itself not to discriminate on the basis of sexual orientation. That means that gays' and lesbians' and their straight friends' and relatives' considerable purchasing power can be brought to bear to convince more corporations to adopt the mark.

The Fair Employment mark website has a lot of parallels to stickK.com. Both are commitment stores that let users quickly enter into binding contracts that cost them nothing as long as they stick to their commitments. And both leave the reffing of the commitments to others. But most importantly, they are both built on the *Field of Dreams* philosophy that "if you build it, they will come." Usually commitment screens and filters are bargained for. But firms might unilaterally signal a commitment to equality—just as Hyundai unilaterally committed to an extended car warranty—as a way of attracting more job applicants or consumers.

One of the most powerful things about the existence of commitment websites is that they can change the conversation. The skeptical negotiator can ask why the other side isn't willing to use it: "If you're not going to stand me up, why aren't you willing to enter into a commitment contract?" But they can also be used proactively and unilaterally by the truly sincere. MillerCoors can prove that it has changed its policies and it is now really committed to equality by publicly committing to equality.

To change the conversation, however, the commitment option must itself be heard. The possibility of committing to equality has to become sufficiently salient so that individuals and firms realize that they have to make a choice. When Jennifer and I launched the Fair Employment mark in 2005, we had ambitious plans to get 100,000 workers covered by its protections within the first year. To be honest, we have not come close to that goal. Even though we have assiduously pitched the idea in blog posts and articles, it never gained sufficient currency that firms felt they even needed to address the question. Yet what would happen if Congress adopted an "opt out" version of ENDA? We tend to think that civil rights laws are mandatory. But imagine a situation in which

Congress merely gave firms the commitment option—by passing a law prohibiting an employer from discriminating on the basis of sexual orientation unless the employer sent the Justice Department a letter saying, "I prefer to retain the legal right to discriminate on the basis of sexual orientation." If the government announced that it would publish a list of the firms who opted out, how many employers could bring themselves to send the letter? Not AT&T or Microsoft. If Congress merely passed an opt-in version of ENDA, I'm pretty sure that my employer, Yale University, would quickly send a letter volunteering to be bound.

Salience really does make a difference. It's one thing for Ian and Jennifer to offer an opt-in version of ENDA. That can pretty easily be ignored. It's another and much more salient option when the government does it. What small progress Jennifer and I have made on licensing our mark has come by bargaining. For example, when a partner from the Watton Law Group, a small law firm in Wisconsin, asked me for help as an economic expert on a case, I conditioned my acceptance on the firm adopting the mark; the firm immediately agreed. The larger point is that even when private commitment options are available, government can play a useful role by helping to put the commitment option on the map.

Government commitment options might even improve the First Amendment. Ever since the landmark 1964 decision in *New York Times v. Sullivan,* the law has immunized media outlets from defamation damages if they recklessly misrepresent facts about a public figure—even if that public figure is truly injured by the misrepresentation. The news media is the only for-profit business that can recklessly injure someone without having to pay tort damages. But the Constitution doesn't prohibit newspapers from volunteering to stand behind the truthfulness of their words. The New York legislature could pass a law allowing newspapers to send, say, the state attorney general a notice committing to compensate victims of negligent misrepresentation.

I wouldn't expect the *National Enquirer* to opt in. But some newspapers might feel a moral duty to compensate people who are injured by their negligence. Other newspapers might commit to compensation as a way to signal to their readers that their reporting is truly "fit to print." The legislature, by making it a more salient option, could instill a healthy kind of commitment competition. Newspapers would feel more pressure to opt in only if their readers valued it.

But as with the ENDA example, the private commitment option already exists. When reporters email me for an interview, I've sometimes responded:

> I would be happy to be interviewed. But I am concerned about the ability of the print media to harm public and non-public figures by negligent misrepresentations of fact without compensating them for their injury. I therefore propose the following contract that I've offered in the past:

———

Agreement to Compensate for Negligent Misrepresentation

In this agreement: _____ shall be referred to as "the reporter"; _____ shall be referred to as "the publication"; and _____ shall be referred to as "the source."

In return for the participation of the source as an interviewee, the publication promises to compensate anyone who is damaged by a factual misrepresentation printed in an article that expressly quotes the source. Compensation for factual misrepresentations is to be measured by the dollar amount required to make the damaged person whole, but in no event shall be less than $100. Damages might be mitigated by timely retractions of the misrepresentation. Anyone explicitly named in the article is an express third-party beneficiary of this contract and thereby has a right to directly sue the publication if it breaches its promise to compensate. The publication and the source intend for this to be a legally binding agreement. The reporter in agreeing to this contract on behalf of the publication represents that the reporter has actual and apparent authority to enter into this contract on behalf of the publication.

To accept this contractual offer (and thereby create a legally binding contract between the publication and the source), please reply to this email with a subject line that states "On behalf of the publication, I accept the Agreement to Compensate for Negligent Misrepresentation."

———

· To be honest, I've used it only when I really didn't want to do the interview. And no reporter has ever accepted my offer. If just one source or just one reader demands misrepresentation compensation, he is a weirdo who is refusing to be interviewed. But if a larger group of readers or sources decides that it is immoral to assist in the production of a for-profit product that refuses to be legally accountable for its negligence, who knows what might transpire?

WHAT'S IN IT FOR ME?

Nowadays, I'm not expecting widespread use of either the Fair Employment mark or the negligent misrepresentation contract. It's just asking too much to expect sources or employees to bargain for contracts that primarily protect other people. Once again, the participation constraint looms large—but here it's even larger, because one group needs to convince a second group to commit to helping a third group. The cynic in me wants to say, Good luck with that.

I initially had a similar response to Dean Karlan's devotion to voting commitments. (You'll hear lots about Dean in the last chapter of this book.) Early on in developing the website, Dean was all het up about creating "voting commitments" for the November 2008 presidential election. He created a special contract where people could commit to vote in the election and then, using publicly available information, Dean could verify whether they had in fact shown up at the polls. If they'd failed to vote, he'd make sure that their credit card would be billed and that emails would be sent to friends they'd designated in advance.

I knew that federal law prohibits paying people to vote for or against particular candidates. But you might be surprised to learn that some states, like California, go further and prohibit paying people to vote in an election. Dean's voting commitments passed legal muster because they didn't pay people to vote. At most, people made a contribution to a charity (or their designated anti-charity) if they didn't vote.

I was skeptical about whether anyone would actually sign up for the contract. I thought it was a neat idea, but I figured there wouldn't be much demand. The hard-core voters don't need a commitment contract to help them vote, and the hard-core nonvoters don't want help in the first place. I thought the commitment contracts would be attractive only to the smaller set of people who wanted to vote but knew they would have trouble getting around to pulling the lever when Election Day actually came. And are these people really going to spend the five minutes necessary to sign up for a contract? I thought not.

Dean, however, has now convinced me that there are good reasons for even hard-core voters to go out of their way to sign commitment contracts. First, signing a voting commitment could reduce how much you are hassled before an election. The current political wisdom is that in the last seventy-two hours before a contested election, a campaign should try to contact its likely supporters seven times. We've been on the receiving end of this deluge—with door hangers, robocalls, letters, emails, and face-to-face canvassing making sure that we're planning to vote.

Signing a commitment contract to vote can serve as a kind of political Do Not Call list. People who've put $100 at risk are much, much more likely to show up to the polls. Campaigns don't need to harass you if you've already made a credible commitment. It hasn't happened yet, but Dean imagines a day when campaigns would happily agree not to call or contact anyone who has signed a commitment contract. Campaigns "would love this," Dean said. "They would know that these are individuals truly committed to voting, so any phone call or visit to them is wasted money. People would love it: fewer harassing phone calls and doorbell rings."

And what helps you is likely to help your candidates, as well. Political scientists have estimated that campaigns can spend $40 per vote on these last-minute contacts. Making a credible commitment to vote is like making a contribution to the candidate for whom you intend to vote.

Those annoying cards and letters that you don't need eat up the real resources of the candidates. Increasingly, candidates can predict how you are going to vote. After you sign a commitment contract, these candidates don't need to waste money and time making sure that you go to the polls; they can spend that money on those who are still at risk of not showing up (or voting for the other candidate).

If you're in a state that lets you vote early, you can generate these benefits just by going ahead and voting for your favored candidate as early as possible. Savvy campaigns don't waste resources contacting people who've already voted. But in traditional jurisdictions, where you have to wait until the first Tuesday in November to vote, voting commitments are a powerful answer to the question "What's in it for me?" because they reduce your hassle and save your candidates money. If voting commitment contracts really acted as Do Not Call lists, even some of the hard-core nonvoters might start signing up—if only to avoid the concentrated pestering that will be directed at those who are still undecided about whether they will vote.

Of course, these benefits are only in the offing if the candidates are listening. And candidates will listen only if enough voters sign up. For the 2008 presidential election, only about three hundred people signed Dean's voting commitment—hardly enough to make it worthwhile for the McCain or Obama campaigns to retool their get-out-the-vote juggernauts. There is still a sizable risk that Dean's idea will not overcome this chicken-and-egg saliency question.

COMMITTED TO CONSERVATION

It's natural to be skeptical that publicly spirited commitments could be both large-scale and voluntary. If we as a nation want to commit to conservation, the only way to do it is through legislation that somehow limits the amount of pollution. For example, under a cap-and-trade system of carbon dioxide permits, firms that emit greenhouse gases would have to purchase carbon dioxide permits as a part of doing business. And Congress would control total emissions by controlling the number of permits. But while Congress debates mandatory cap-and-trade systems, a voluntary-commitment regime with little fanfare now governs the production of about 4 percent of our country's greenhouse gas emissions.

The Chicago Climate Exchange is an object lesson for those who think that publicly spirited commitments are nothing more than pipe dreams. More than 350 firms have entered into binding and independently verified contracts to reduce greenhouse emissions by 1 percent a year. Companies like Ford, DuPont, and Motorola, as well cities (Chicago and Oakland) and colleges (University of California) have all agreed to multiyear contracts with financial penalties if they fail. If they surpass their goal, they end up with permits they can sell to others. If they fail, then under the terms of the commitment contract, they must buy permits from others.

Why in the world did for-profit companies like DuPont take on the risk of an monetary liability if they fail to conserve? Some firms want to give themselves an extra financial incentive to conserve. Others may benefit by sending a credible signal. The Chicago Climate Exchange emphasizes that members "gain leadership recognition by taking early, credible and binding action" and "prove concrete action on climate change to shareholders, rating agencies, customers and citizens."

But the economist in me loves that the commitment contracts are structured to entice firms for the most traditional economic rationale of all: to make money. They can sell their excess credits to those who fail to meet their goal. And even though the goal is different, the climate exchange agreement is really quite analogous to the group weight-loss agreements. Here the goal is to lose tons of carbon dioxide instead of pounds of fat, but the basic idea is still the same. In both cases, those who fail promise to pay those who succeed.

Utilities could offer the same opportunity to individual households. You could contract to reduce your home energy consumption by 1 percent a year for each of the next ten years. When you beat that target, you'd get permits to sell. When you miss, you'd pay a penalty by buying unused permits from others. As a result, your incremental price of fuel would go up. Every extra BTU you use would mean fewer permits to sell or more to buy.

This incremental price shift is like a self-imposed tax on energy consumption. It creates an incentive to conserve. But it's different from an ordinary tax in that you may not have to pay anything. If you make your target, you won't owe a dime. Thus, the scheme gives you an incentive without hitting you over the head.

Creating a real BTU exchange where credits are bought and sold might be cumbersome. But your utility could set it up and do all the trading for you. When you sign up, the utility company would give you permits equal to 99 percent of your previous year's consumption. To the extent that you beat that 99 percent target, you'd get credit for the value of your extra permits, with a minimum price of 5 cents per kilowatt-hour. Five cents is about how much utilities are already spending, per kilowatt-hour saved, on programs to help consumers conserve energy, and it's enough to provide a big incentive. It's about 75 percent of the average price for electricity in the United States, not including transmission and distribution costs. When you missed the target, the utility would buy the extra permits and bill you, up to a maximum price of 10 cents per kilowatt-hour. Utilities should be allowed to buy renewable energy credits from one another, too, if it makes for a cheaper way of obtaining them. So someone turning off an air conditioner in San Diego might be paid a fee by an energy squanderer in Atlanta.

Many consumers already buy offsets for their auto emissions or jet travel. That's noble, but it doesn't reward people who insulate their houses or find other ways to conserve energy at home. Joining a "Consumer Climate Exchange" would offer them specific targets and financial rewards.

There's still an important role for government here. But it's very different from the traditional command-and-control model of regulation. Instead, the model is government as facilitator, government as verifier, and maybe government as backstop. If every household managed to beat its goal, the utility company would end up in the hole, buying more excess credits at 5 cents per kilowatt-hour than it could sell. The government might stand ready as a purchaser of last resort, to ensure that any excess credits would be bought for a price of at least 5 cents.

At the end of the day, I'm still a fan of some command-and-control regulation. Like most economists, I favor increasing the gas tax as a collective commitment to promote energy independence and conservation. Gas taxes also save lives, as highway fatalities decrease with less driving. However, the stark political truth is that neither political party is going anywhere near a traditional gas tax.

But as with the Chicago Climate Exchange, it is possible to structure a voluntary gas tax that compensates citizens for committing to conservation. It's hard for even the staunchest libertarian tax critic to be against giving people the option of committing to be taxed for the good of the country.

Voluntary taxation seems like an oxymoron. No sane person would ever volunteer to be subjected to a tax. Yet about half the cost of World War II was financed by voluntary purchases of war bonds, which had a sort of tax built in because they paid below-market returns. Buying them was the patriotic thing to do. Bond rallies with stars like Rita Hayworth and Bette Davis generated mass support for "the greatest investment on earth."

The Obama administration could use something like this to help win our nation's fight for energy independence. An "independence bond" would pay you a lump sum of cash today if you opted to pay the government at the end of the year an extra tax for every gallon of gas you used in the interim. Those who ended up driving their cars less would end up making money from the deal. And that's the point: to reduce our dependence on foreign oil.

Here's the recipe: The government would offer a $500 advance-tax rebate each year for every car if the car owner chooses to sign up for the tax. In return, the car owner would agree to pay an extra $1 for each gallon of gas she buys. The actual tax paid would be based on miles driven and the car's fuel economy rating. Thus, a Chevy Impala rated at 19 miles per gallon might be charged $5.26 for each 100 miles, while a Prius rated at 46 miles per gallon would be charged $2.17 per 100 miles.

For cars with average fuel efficiency (22.4 miles per gallon), you'd break even if you drove 11,200 miles a year. People who already drive their cars less or who drive fuel-efficient cars would be particularly likely to opt for the independence bonds. But even these folks would have a strong economic incentive to reduce their driving. (Since most people can't or won't cut their consumption by 50 percent, it might be better to pay $250 up front and then charge people only an extra $1 per gallon for their use above 250 gallons. That would cut down on the cost of the program while still providing all the right incentives.)

Because the plan is optional, high-mileage drivers and businesses that can't afford the extra cost would be unlikely to sign up. But the history of war bonds suggests that, if marketed properly as an act of patriotism, independence bonds might appeal to a large swath of Americans. Cars that were committed to conservation would be eligible to display decals showing their patriotism in the fight for energy independence. The decals might also authorize the use of highway lanes now reserved for buses and carpoolers.

Unlike the Chicago Climate Exchange, which compensates people after the fact for conservation, independence bonds compensate you in advance. Instead of a rebate, think of it as a prebate. From a behavioral perspective, the offer of immediate cash in hand is a powerful lure. Like the Hurman auction, the independence bond pays people in advance to take on the risk of a future forfeiture. No one is forced to accept, but the offer is likely to be particularly attractive to those who have willpower problems because they hyperbolically discount the future. Here's one time where impatience, environmentalism, and even economic stimulus all go hand in hand.

In the spring of 2008, Chrysler promised to subsidize the price of gas down to $2.99 a gallon for a year for anyone who bought one of its cars. Back in 2006, Hummer and GM promised that you wouldn't pay more than $1.99 a gallon for a year. These promotions tapped into the same demand for lower gas prices that fueled Hillary Clinton's and John McCain's proposals for a summer gas-tax holiday. The problem with all these plans is that they subsidized people to drive more.

Conservation bonds are just the reverse: participants are paid today for accepting higher gas prices in the future. A company like Toyota might even offer to match the government plan: Prius owners could get a $1,000 rebate if they promise to pay Toyota an extra 2.2 cents for every mile driven the first year. While car companies might use conservation rebates to attract green customers, a call by our leaders to voluntarily embrace credible conservation would engage an even larger number of Americans.

The government's role would again be crucial—not just in whipping up public support and bankrolling the prebates, but in making sure that people who took the money up front followed through and paid the additional tax. People who claimed the rebate would need to have their odometers checked once a year to calculate the amount of tax owed. It's difficult to roll back an electronic odometer, and odometer readings would be particularly easy in the thirty-three states that require periodic vehicle inspections.

It's fine to be skeptical. The smart money is almost always on congressional inaction. Oil companies (and even just retailers who rely on car traffic for sales) have powerful incentives to oppose even voluntary gas taxes. But in the future, if mandatory taxation starts to gain political traction (but still falls short of enactment), a voluntary program might pick up enough votes to pass.

A DO NOT PAY LIST

In December 2006, Nick Del Giudice had the kind of night gamblers dream of. At the Grand Victoria Casino in Elgin, Illinois, he caught fire at a craps table and won more than $20,000. But his euphoric rush came crashing down when he went to cash in. The casino not only refused to pay him but added insult to injury by charging him with trespassing. "It's ridiculous," he told the *Chicago Tribune*. "The one time I win big, they confiscate it."

Nick's problem is that back in 1998 he had enrolled in the voluntary "self-exclusion program" run by the state. The Grand Victoria personnel didn't pay him because they couldn't. Under the law, casinos are obligated to check winners' names against an electronic state database before making any payouts of more than $1,200. When they ran Nick's name for tax purposes, it popped up on the self-exclusion list.

The casino had every incentive to follow through on its side of the bargain. If a self-excluded gambler hits a million-dollar jackpot, the casino can avoid having to pay him (and can keep the money he initially wagered). What a buzz kill. But of course, that's the point. If you know that you won't ever be able to collect a big payout, gambling becomes a lose-lose situation—especially now that casinos also have a legal obligation to charge the self-excluded with misdemeanor trespass. Casino employees are even paid a $250 bounty if they catch a self-excluded gambler on their premises.

The Illinois program is a perfect example of how commitment options can change the conversation. Nick signed up for the program as a way of convincing his girlfriend that he was serious about giving up gambling. Repentant gamblers in Illinois have something tangible they can offer to their loved ones. And just as important, loved ones have something tangible they can demand. Tens of thousands of people have signed up with government self-exclusion programs, which now exist in seven states (as well as around the world in Australia, Canada, France, the Netherlands, Poland, South Africa, and Switzerland). And the number one reason people give for signing up is to save their marriages. For Nick, making the commitment led to marriage. He married his girlfriend, had a kid, and stopped going to the riverboat casinos for about seven years.

But then he started backsliding on his commitment—going to play about once a month. In September 2005, he was charged with trespassing at the Empress Casino in Joliet. Voluntary self-exclusion is not a panacea. In Illinois, where about five thousand people have tied their hands, there have been more than one thousand violations and more than $500,000 forfeited and donated by the state to anti-gambling charities. But overall, surveys show that the program is having a deterrent effect. It's working so well, in fact, that the state has created an analogous Do Not Pay list for the lottery.

We're used to thinking of government helping promote commitments by enforcing private contracts. But here government is helping problem gamblers by mandating that certain gambling contracts not be enforced. Letting the casino or lottery avoid paying can usefully take all the fun out of gambling. Where's the flutter if you really don't have the possibility of hitting it big? Problem gamblers can enter into these contracts with individual casinos—and, in fact, Harrah's has been a leader in providing self-exclusion options. But the government is uniquely positioned as a one-stop provider of the Do Not Pay commitments. The only problem with the Illinois list is that it doesn't stop Nick from driving out of state to gamble. The federal government might help by providing a list that governs all fifty states. Casinos in other countries and illegal alternatives (Nick's local bookie and online sites) would still beckon. But even leaky buckets can carry a lot of water.

The nonenforcement option might also help in other contexts. A Do Not Pay list might even be applied to the problem of payday loans. Dean Karlan imagines a guy who knows he's particularly weak on Fridays; he gets a payday loan "and then blows it for the weekend on parties and booze and strip bars." If he signs up for a Do Not Pay list that eliminates his duty to pay back any payday loan taken out on a Friday, he is not going to be lent much money on that day. The lenders' computers will be sure to block the attempt, because the lender wants to be paid back. This example also shows that Do Not commitments need not be all or nothing. Some people might commit to no loans on particular days or to a limit on how much they can borrow. But the big idea is that government can create value by giving people (and their loved ones) the option of tying their hands.

It's natural to think of commitment contracts as a tool to strengthen our resolve. Hyperbolic discounters like me can use commitments to win the battle of the bulge or to quit smoking. But commitments can also be used to communicate. Government can enhance our freedom of speech by giving us the opportunity to speak credibly, to stand behind our promises and predictions. First Amendment scholars will tell you that freedom of speech also concerns the right of listeners. Creating opportunities for listeners to separate the credible from the incredible can improve the marketplace of ideas and move us away from a world where false and inaccurate talk is cheap.

7

Antonio's Problems

If you repay me not...,
 . . . let the forfeit
 Be nominated for an equal pound
 Of your fair flesh, to be cut off and taken
 In what part of your body pleaseth me.

In *The Merchant of Venice*, with these fateful words Shylock secures from the good merchant Antonio one of the most famous and tragic commitments in all of English literature. Antonio is so sure that his bond is secure that he is willing to enter into this agreement. After all, he has three ships at sea—all due home before the bond comes due, and any one of which will provide him with the money to pay Shylock. And Antonio so strongly desires to help his young friend Bassanio (to woo the beautiful Belmont heiress Portia) that Antonio is willing to risk his life should he fail.

The narrative is fiction, but the Bard captures a particular concern with commitments as signals and commitments as screens. Before, we've seen how naïveté can dissuade people from being willing to enter into commitment contracts that would really help them. The naïf believes she isn't going to have a problem losing weight in the future, and thus thinks she doesn't need to go through the whole rigmarole of entering into a commitment contract. But Antonio's tale shows how the problem of naïveté can be flipped on its head when someone else demands a commitment. Now we see the unsophisticated promisor who's willing to commit to excessive damages because he underestimates the future difficulty that he is going to have following through.

Credit card companies and movie rental stores exploit cognitive weaknesses when they jack up the price of late fees that consumers don't think they are ever going to pay. Why should you worry about the interest rate on your credit card if you always pay it off on time? This reasoning makes the card issuers millions of dollars because once every year or two even a conscientious consumer forgets to pay the monthly bill on time and is socked with the interest penalty.

Antonio's willingness to offer a pound of flesh might also be driven by a perverse kind of signaling competition. Economists have shown that potential borrowers competing for scarce funds are sometimes willing to agree to "arm-breaking" contracts, inefficiently over-signaling lenders by agreeing to excessive penalties for failure to pay. Borrowers with good prospects of repaying agree to arm-breaking contracts to distinguish themselves from borrowers with poorer prospects—who would be willing to agree to only more moderate penalties. The willingness of bad borrowers to try to imitate the commitments of good borrowers forces the good borrowers to offer excessive arm-breaking penalties. In these mathematical models, the world can be a better place if the law prohibits the option of excessive penalties.

The common law of contracts does just this. As in Shakespeare's Venice, courts will not enforce pound-of-flesh or arm-breaking penalties. Even monetary penalties are closely scrutinized. Common-law courts refused to enforce *in terrorem* provisions, which evoke terror in the promisor contemplating breach. The late, great contract's treatise writer Allan Farnsworth summed up the legal antipathy toward monetary penalties by saying, "It is a fundamental tenet of the law of contract remedies that an injured party should not be put in a better position than had the contract been performed." Liquidated damages provisions, which set out financial penalties in advance, will be enforced only if they are not grossly disproportionate to the actual injury caused by a promisor's breach.

When a romantic tries to do a good thing and fails, they give him
a medal. When a pragmatist succeeds, they wish him in hell.

STEPHEN KING, "QUITTERS, INC."

In literature, the use of commitment contracts with severe penalties is associated with evil. In the Harry Potter series, it is the thoroughly malevolent Bellatrix Lestrange who demands that Severus Snape signal his loyalty to the dark lord Voldemort by making an "unbreakable vow," which kills the promisor if it is broken. In contrast, the saintly Dumbledore demands at times that Harry promise to follow his instructions, but he never demands an unbreakable vow. The mischievous Fred and George Weasley were severely chastised by their parents when they came close to inducing their five-year-old brother, Ron, to make an unbreakable vow. Good guys don't demand arm-breaking penalties.

Stephen King's short story "Quitters, Inc." tells the fanciful tale of mobsters who scare smokers straight by threatening a punishment schedule with ten steps, each one of increasing severity. The first step is thirty seconds of a painful electric shock delivered to a loved one, and later steps include literal arm breaking. It's easy to react with revulsion to this ghoulish scheme. Most egregiously, the company overcomes the participation constraint by simply not telling the clients at the time of contracting what's in store for them if they fail, so the clients don't know that they're putting their lives and limbs at risk. But what makes the story so compelling is the transformative power of the commitments to improve the quality of the clients' lives. King emphasizes that the clients become not only healthier but happier. They win the long-sought-after promotion. Even the loved ones who bear the brunt of their failure are thankful for the changes wrought by the commitment. In the fictional account, 40 percent of clients never experience a single punishment, and only 10 percent reach the fourth punishment, which involves something more severe than electric shock. A mere 2 percent are "the unregenerate" and face the ultimate punishment. One way or another, Quitters, Inc., guarantees that you will not continue smoking.

But here's a hypothetical reaction to these hypothetical statistics: Should the law stop people from entering into a Quitters contract if they know what they are getting into—including the probability of failure? Even though the commitments end up killing 2 percent of the clients, the technique as a whole is so successful at stopping people from smoking cancer sticks that on net it might very well be saving lives.

Of course, there are lots of reasons to refuse enforcement. It almost goes without saying that the engine of the state ought not to be used to facilitate the killing of people who do not want to die. Just as we don't let people sell themselves into slavery, the state should not enforce suicide pacts. This is what the Declaration of Independence calls an "unalienable right." Living in a world where such things are possible, in some ways diminishes us all. And my willingness even to entertain the question probably reveals some deep deficiency in my moral compass.

Luckily, stickK steers far, far away from such unsavory waters. We limit our users' stakes, the amount they put at risk on a contract, to less than $10,000. And our terms and conditions require users to represent that "the total of all Commitment Stakes authorized by Client is less than 10 percent of Client's annual income." But we were concerned about more than just the size of the stakes. Dean Karlan, Jordan Goldberg, and I (the cofounders, whom you'll learn more about in the next chapter) thought long and hard about how to limit the commitment service so that it would not become a force for bad in the world. In my day job I teach contracts at Yale Law School—so it was particularly rewarding for me to get a first crack at drafting the language for the commitment contracts. At heart, our product is nothing but a contract. You'll soon learn that Dean and Jordan's contributions to stickK far outstripped mine. But composing these initial drafts (which were substantially rewritten and rewritten again) is probably my most substantive contribution to the enterprise.

We also put noncontractual protections in place. We added technological barriers to stop people from committing to lose too much weight or to lose it too fast. We block weight-loss contracts where the ultimate weight produces a BMI of less than 18.5 or where the rate of weight loss is more than two pounds a week. Our contract terms also exclude clients who have "been diagnosed with an active condition of anorexia or bulimia nervosa, respiratory disease, heart condition or any other condition that would make the Commitment unhealthful." And we have both automated and human scans of all custom-designed contracts to make sure that they are not commitments to break the law or further some antisocial end. Commitments to quit using drugs are fine; commitments to use more cocaine or to mistreat your neighbor are not. The good news is that to date we've never had to cancel a contract because it would hurt others and only a few times have had to remove commitments to try to reach anorexic BMIs.

Most importantly, we added a medical excuse to all contracts to protect people who enter into a commitment imprudently, or where something changes to make what was once a reasonable commitment no longer reasonable. If a woman who committed to losing a lot of weight later became pregnant, we wanted to make sure for both moral and legal reasons that she had a simple and clear mechanism for getting out of the contract. Section 9.2 of the commitment contract contains this escape hatch:

This Commitment Contract may be cancelled by Client if and only if a licensed physician faxes a completed and signed copy of the stickK "Medical Excuse" form . . . to the fax number indicated on the form.

The language is crucial because it determines the amount of flexibility clients will have in avoiding their commitments. We could have made it broader and let a client out of any commitment where the client asserted that it "would be detrimental to my health to continue to be bound by this commitment." But going that far would give every Odysseus a way to unfetter his own hands. Alternatively, we could have relied on our own discretion to decide when to cut people a break. But the essence of stickK is not just its contracts but its reputation for following through and enforcing commitments. We didn't want to be put in the position where we were making decisions about what is "healthful." Instead, we thought that asking for a signed verification by a licensed physician was the appropriate compromise. To be sure, we realized that a user who approaches enough physicians is sure to find one who will attest to just about anything. But forcing the user to go to the effort in and of itself provides some friction to make the commitment at least a bit, shall we say, stickier. And it was essential that we made it easy for people with good medical reasons (including psychological distress) to be able to get out of their contracts.

In providing for medical excuses, stickK is in fact holding itself to a much higher ethical standard than many other contractual commitments. Something that is as ubiquitous as a home mortgage is, after all, a kind of commitment contract: as mentioned above, you commit to paying back your loan or risk losing your home. But if you lose your job or a loved one falls ill, trying to follow through on your mortgage commitment can create a lot more psychological strain than the prospect of losing a few hundred dollars on a stickK commitment. As I write this passage, in (what we hope to be) the aftermath of the 2008–9 housing crash, this concern is hardly hypothetical—as reports of "foreclosure-related" suicides have become all too frequent.

But unlike stickK, your mortgage company—or, for that matter, your telephone, credit card, or utility company—won't let you out of your commitment to pay if a physician says that payment would be deleterious to your health. Some might respond that these companies, unlike stickK, shell out costly goods and services that induce customers' promises to pay and that it would destroy their ability to cover costs if they had to excuse repayments that would cause undue mental distress. But I hope by now you realize that providing a credible commitment mechanism can also be a valuable service. I assure you that providing this service (maintaining the website, paying salaries, and so on) is also costly. And our ability to cover costs would also be jeopardized by letting people back out of their commitments. The common law of contracts has already created limited opportunities for letting promisors out of contracts when performance becomes impracticable or there has been a frustration of purpose. Out of an abundance of caution, we decided to go further and create a clear, additional exit option to make sure that commitments didn't end up hurting the health of customers.

HOW DO YOU ACCENT ATTRIBUTES?

It may seem excessive for me to risk $26,000 in a single year in order to keep my weight down. But that's chicken feed compared to the $2 million Curt Schilling put on the line in 2007 to keep his weight in check. In November of that year, Schilling signed a one-year contract to keep playing for the Boston Red Sox, which guaranteed him an $8 million base salary. He could also, however, make $6 million more in incentives. He would earn an extra $1 million if he received a single vote for the Cy Young award, and an additional $3 million if he pitched 200 innings. But most bizarrely, the contract called for "six random monthly weigh-ins," at each of which Schilling could earn $333,333 if he kept his weight below 230 pounds.

"[The weight-loss incentive] came off the heels of the '07 spring training," Schilling said, "when I showed up and was heavier than I wanted to be, and people went nuts thinking that I was out of shape." After the season was over, Schilling and the Red Sox were looking for ways to structure a contract with a lower guaranteed salary but with incentives that would give him an opportunity to maintain his earnings close to what they had been—$13 million. "I inserted the weigh in clause in the 2nd round of offers," Schilling wrote on his blog 38 Pitches. "Given the mistakes I made last winter and into Spring Training I needed to show them I recognized that, and understood the importance of it."

But when I interviewed Schilling in 2009, after he had retired from baseball, he rejected the idea that a $2 million weight-loss incentive would have affected his performance. "I think that they felt like that was going to be some sort of incentive for me to maintain weight or whatever but the incentive for me is winning ball games."

Schilling already had a lot of experience with the pay-for-performance idea. For more than a decade, he'd helped fight amyotrophic lateral sclerosis (a.k.a. Lou Gehrig's disease) with his Curt's Pitch for ALS Program, which allowed fans to pledge money for every strikeout he threw. And Schilling, who threw more than 3,000 strikes, raised over $6 million. But as proud as he is of helping fight ALS, he still rejects the notion that the promotion spurred his performance. "I can't ever remember a second of my life when I was thinking, 'Wow, I really need to be as good as I can be to get this incentive," he explained. "[When] we won the World Series . . . I did not even think about [the incentive] until the next morning. It didn't hit me that that actually triggered that incentive."

It's hard to believe that Schilling would be completely impervious to the impact of a $2 million weight-loss incentive. Even if his focus on the field is all about winning games, his off-season preparation and even his training during the season could have been influenced by all those dollars at stake. Then again, this is the same guy who was twice willing to pitch on a newly sutured ankle as his sock became soaked with blood. So I'm not thinking that money was foremost in his mind when he willed the Red Sox past the Yankees in Game 6 of the 2004 ALCS.

But it's still amazing how natural it is to chalk up life's bad outcomes to circumstances beyond our control. "It is one of those genetic things," Schilling said. "My body type—I am built like my dad. I carry all of my weight in my upper body. I don't have very big legs. So, I carry weight differently." He wrote on his blog, "I was completely broad sided by the fact that your body doesn't act/react the same way as you get older. Even after being told that for the first 39 years of my life."

The reality is that Schilling's increased weight was caused by a variety of factors, including his genetics and how much he chose to eat. The extent to which a bad outcome is caused by your own bad choices is an empirical question. But psychologists have discovered a truly disturbing tendency with the wonderful name "the fundamental attribution error" that makes it hard for us to suss out the true causes of misfortune. When something bad happens to others, we tend to think that it's their fault. When something bad happens to us, we tend to think it's not our fault— that it happened because of something beyond our control. When we see someone else shooting baskets badly in a dark gym, we tend to think it's because he's a bad shot. When we shoot badly in the same gym, we tend to think it's because there's not enough light. We underplay the importance of the context when we are observing others; we overplay the importance of context when we are observing ourselves.

This is a dangerous dynamic when it comes to negotiating commitment and incentive contracts. The Red Sox are wont to overemphasize Schilling's poor eating choices in explaining why he came into spring training at a high weight, while Schilling is likely to overemphasize his poor genetic luck. The fundamental attribution error can lead others to demand too many commitments from us.

And conditioning payment on outcomes that really are beyond our control can be downright insulting. At about the same time that the Red Sox extracted a $2 million weight incentive from Schilling, Joe Torre balked at an incentive-laden, one-year offer to continue managing the New York Yankees. The $1 million bonus for getting the team back to the World Series particularly rankled him. Psychologist Barry Schwartz, the author of *The Paradox of Choice,* chided the Yankees for insisting on the bonus, observing that "the offer of a bonus implies that without it, the employee would just be mailing it in." In earlier chapters, we've worried that financial incentives can undermine the intrinsic motives and extrinsic social norms that drive performance. But the Torre example shows that merely offering an incentive can be destructive.

In the end, we will never know whether either the Schilling or the Torre incentive would have worked. Torre went west to coach the Dodgers, while Schilling wound up missing all of the 2008 season because of a shoulder injury. Curt never was weighed, because he offered to modify that part of the contract. He said, "I could have sat on the deal in the postseason with the weight clause and collected the money, but that is not me."

A PLAN FOR ATTAINING "MORAL PERFECTION"

It's natural to see a clause that ties $2 million to pounds of flesh as another Antonio contract that simply puts too much at stake. But an independent concern is whether it's wrongheaded to incentivize too many aspects of your life. At one extreme, we could imagine a grim world where one resolved to perfect every aspect of one's behavior and backed up the resolution with strict accounting and intervention. In fact, this experiment has already been run by none other than our most protean of founding fathers, Benjamin Franklin.

As a young man, Franklin "conceiv'd the bold and arduous project of arriving at moral perfection." He wrote, "I wish'd to live without committing any fault at any time. . . . As I knew, or thought I knew, what was right and wrong, I did not see why I might not always do the one and avoid the other."

Franklin was a man of seemingly unbounded abilities. But perfection was not so easily obtained. He admitted:

I soon found I had undertaken a task of more difficulty than I had imagined. While my care was employ'd in guarding against one fault, I was often surprised by another; habit took the advantage of inattention; inclination was sometimes too strong for reason. I concluded, at length, that the mere speculative conviction that it was our interest to be completely virtuous, was not sufficient to prevent our slipping.

Franklin's experience of "guarding against one fault" only to be "surprised by another" is still with us today. When I commit to cut down on my TV watching, I see a pronounced increase in my (mis)use of the Internet. This is analogous to the increasing recognition of "addiction transfer." The Betty Ford Center reports that "about 25 percent of alcoholics who relapse switch to a new drug"—in some cases, opiates. We've heard of weaning addicts away from heroin with methadone replacement therapy.

But attempts to wean addicts away from alcohol can sometimes spur substitution toward other destructive addictions.

The substitution problem seems to be particularly pronounced after obese patients have their stomach capacity reduced with bariatric surgery. Therapists at some weight-loss centers now estimate that between 20 and 30 percent of bariatric-surgery patients pick up other compulsive disorders after surgery—including smoking, gambling, and even compulsive shopping. Alcohol abuse is a particular problem, possibly because alcohol is high in calories and can pass quickly through the surgically reduced stomach. To date, the studies are small and preliminary, but with more than 100,000 new bariatric patients a year (and growing), information about addiction transference has become a standard part of pre-op counseling.

Before committing to stop one bad habit, we should consider the chances that success will promote other bad habits. Even with some substitution, the game can still be worth the candle. People who successfully quit smoking often put on a few pounds but still add years to their expected longevity. And there's always the possibility of taking up separate arms against the substitute behaviors. In "Quitters, Inc.," Stephen King imagines that 73 percent of clients who quit smoking start gaining weight. His goons helpfully ensure that they stay trim by threatening to cut off their spouse's little finger if their weight ever drifts above a magic number. This mandatory added service is not disclosed to clients until they are well into the smoking treatment.

Benjamin Franklin took on the burden of multiple commitments of his own volition. When the inventor of the public library found that he was failing in his quest for moral perfection, he constructed a list of thirteen virtues—temperance, silence, order, resolution, frugality, industry, sincerity, justice, moderation, cleanliness, tranquillity, chastity, and humility ("imitate Jesus and Socrates")—all of which he hoped, over time, to master. His hubris is simply remarkable. *The New Yorker* once ran a cartoon that showed James Joyce's refrigerator, on which was stuck a "to do" list that mixed together such ordinary tasks as "call bank" and "dry cleaner" with "forge in the smithy of my soul the uncreated conscience of my race." But Franklin, on a personal level, was attempting something no less audacious. And while the point of the cartoon is to lampoon the idea of a refrigerator list with such abstract goals, Franklin—ever the pragmatist—kept a daily ledger of his progress:

I made a little book, in which I allotted a page for each of the virtues. I rul'd each page with red ink, so as to have seven columns, one for each day of the week, marking each column with a letter for the day. I cross'd these columns with thirteen red lines, marking the beginning of each line with the first letter of one of the virtues, on which line and in its proper column, I might mark by a little black spot, every fault I found upon examination to have been committed respecting that virtue upon that day.

I've been slightly embarrassed to make daily spreadsheet entries on five attributes of my own life—but long before me, Franklin hand-drew his own spreadsheet to keep track of thirteen dimensions of his life. And some two hundred years before Skinner, Franklin undertook something that is at least a step toward a "shaping" intervention. After his initial failure at across-the-board attention, Franklin chose to divide and conquer, attacking his vices individually (all the while keeping track of the degree of backsliding on other dimensions):

I determined to give a week's strict attention to each of the virtues successively. Thus, in the first week, my great guard was to avoid even the least offence against Temperance, leaving the other virtues to their ordinary chance, only marking every evening the faults of the day. Thus, if in the first week I could keep my first line, marked T, clear of spots, I suppos'd the habit of that virtue so much strengthen'd and its opposite weaken'd, that I might venture extending my attention to include the next, and for the following week keep both lines clear of spots. Proceeding thus to the last, I could go thro' a course compleat in thirteen weeks, and four courses in a year.

Ben kept at it for several years, but never did end up with a clean book. The big question is whether his efforts were worth the cost.

SHOULD PRESIDENT OBAMA KEEP SMOKING?

Of all Franklin's goals, trying to maintain order particularly vexed him and made him question his undertaking:

This [virtue] cost me so much painful attention, and my faults
in it vexed me so much, and I made so little progress in amendment, and had such frequent relapses, that I was almost ready to
give up the attempt, and content myself with a faulty character
in that respect.

Franklin likened his endeavor to that of a man who tried to convince
a smith to grind an ax so that the whole of its surface would be "as bright
as the edge." The smith consented to grind it bright for him if the man
would but turn the grindstone. After turning the wheel at length with
only small effect, the man tried to break off, saying, "I think I like a
speckled ax best." Who knows how many more conveniences we might
have today if the inventor of the lightning rod, the library, and the
Franklin stove had spent a bit less time trying to order his papers.

I think of Franklin's fruitless efforts when I selfishly consider
whether now is the right time for President Obama to kick his smoking
habit. A poorly kept secret is that our forty-fourth president considers
himself only "95 percent cured" and has admitted that there are still
times he "mess[es] up" and smokes. At a time when our country faces
extraordinary challenges at home and abroad, is it prudent for our com-
mander in chief to devote his limited energies to kicking the habit—even
if smoking means he can expect to die sooner?

I'm worried because there is increasing evidence that self-control is
a limited resource that can be depleted if it is overused. Which is harder
to resist: eating a warm chocolate-chip cookie or eating a radish? You
don't need a PhD to know that for most people, almost anything chocolate is harder to resist. But Florida State University psychologist Roy
Baumeister used people's weakness for chocolate to test whether exerting more self-control would make it harder for them to follow through
on other tasks. Baumeister found that students who had to resist eating
chocolate-chip cookies were likely to give up twice as fast when they
were then asked to solve an unsolvable puzzle relative to students who
had to resist eating a radish.

A single cookie study isn't enough to change my mind. But Baumeister and a cadre of coauthors have done dozens of studies showing that
"ego depletion" can make it more difficult to succeed in subsequent tasks
that require self-control. For example, it's hard for me not to laugh at
Robin Williams's comedy or to resist crying after seeing Deborah Winger

say good-bye to her kids in *Terms of Endearment*. Baumeister figures it's probably hard for other people, too. He showed these clips of Williams and Winger to a group of subjects and then asked them to try to solve as many anagrams as they could in ten minutes. Half the subjects were asked to exercise some self-control and not display any emotion when watching the film clips. The other half were told "to let their emotions flow" while watching the film. Once again, Baumeister found that people who had to exert self-control during one task (here, suppressing emotion) had greater trouble exerting self-control (here, diligently following through) during another. People who suppressed the urge to laugh and cry solved, on average, 4.9 anagram problems, while those who were not constrained solved 7.3 problems.

Ego depletion might not just undermine problem-solving persistence; it might undermine honesty as well. In another experiment, Baumeister depleted subjects' self-regulation capacity by having them write an essay without using words containing certain letters. He then tested their honesty by paying them twenty-five cents for every correct question on a self-graded math quiz. Half the students had an essay task that required more self-regulation, because their essays couldn't use words with the common letters *a* and *n*. The other half had the much easier task of writing an essay that avoided words with the letters *x* and *z*. The more depleted students were less honest when they claimed their prizes. Students who had completed the more arduous (no *a* or *n*) task claimed 60 percent more math prizes than the students who had completed the less arduous (no *x* or *z*) task. And because of random assignment, we can be confident that the depleted students were not innately more adept at math problems than the less depleted students.

These studies counsel against taking on too many commitments at once. Ben Franklin may have found himself slipping up on sincerity and humility when he was concentrating on order because he was misallocating a scarce resource, his willpower. Ego depletion has a lot to say about why even successful attempts to control one's bad habits often lead to substitution toward other, seemingly unrelated addictions. With so much else demanded of him, now may not be the best time for Obama to try to kick the habit.

Or maybe, if he does, he might consider drinking some lemonade. The idea of a weakened ego as less able to reign in the baser urges of the id is a powerful, if slightly dated metaphor in this post-Freudian world.

But Baumeister has more recently shown that something as concrete as blood sugar levels may underlie the ego-depletion metaphor. As he puts it, "acts of self-control reduce blood glucose levels." He asked dozens of subjects to watch a six-minute video of a woman talking, while in the corner of the screen common words (e.g., "hat," "hair") individually came onto the screen for ten seconds each. The subjects' glucose levels were measured before and after each of the screenings. To manipulate ego depletion, the experimenter at random told half of the subjects to ignore the words and focus their attention solely on the woman, while the other subjects (the control group) were told to watch the video as they would normally. The control group showed no decline in glucose after watching the video, but for the group told to control their attention, glucose levels fell on average by 6 percent. Just six minutes of self-control is enough to dampen your glucose levels. Moreover, those subjects with depleted glucose had trouble exerting subsequent self-control. In another experiment, Baumeister asked the same two groups to complete eighty Stroop tasks. The Stroop test (named after John Ridley Stroop) is one of the best-known and elegant tests of self-control. Subjects are merely asked to say out loud the color of the letters they are shown. What makes it difficult is that the letters spell out a color that is different from the color of the letters used to make the word. For example, if you were shown this word:

GREEN

you should say, "Black," because that is the color of the font (as it is for all the words in this book). It requires self-control to turn off our reading impulse and respond solely to the color of the actual letters. Sure enough, Baumeister found that the subjects who were ego- and glucose-depleted from trying to ignore the words on the TV screen had a harder time than the randomly selected control group in saying the names of the colors they had learned in kindergarten.

So I wasn't kidding when I suggested that a glass of lemonade might help. Baumeister reran the experiment, but started by having all the subjects drink a glass of lemonade. Unbeknownst to them, however, half of the glasses were sweetened with Splenda artificial sweetener and had 0 calories, while half the glasses were sweetened with real sugar and delivered 140 calories of glucose. In one of the more amazing experimental results in this book, Baumeister found that simply drinking a glass of real lemonade eliminated the depletion effect. The subjects who drank the Splenda-laced placebo continued to have trouble controlling their urge to read the letters instead of identifying the colors. But the subjects who drank some sugar water were able to hit their nondepleted levels of accuracy. No wonder we eat a bit more when we are trying to quit smoking—we may be indirectly trying to increase our resolve by boosting our blood sugar.

Then again, a bit of ego depletion may be good for the soul. Baumeister and his colleagues are beginning to think that an individual's capacity for self-control is like a muscle that can be strengthened with regular exercise. Even though muscles exhibit reduced strength immediately after exercise, the same muscles, after recovering, can become stronger. Researchers have started to ask if the same thing might be true of our capacity for self-control. By now it won't surprise you to learn that subjects who were asked not to think about a white bear exhibited less stick-to-itiveness when subsequently asked to see how long they could squeeze a hand grip. This is just another example of ego depletion. But the depletion effect went away for subjects who had for the previous two weeks been assigned the task of improving their posture—told to sit up straight whenever they thought about it. Exercising self-control in one area of your life may give you more resources to exercise it in other areas, even when you would otherwise be depleted.

One of Baumeister's coauthors, Megan Oaten, an Australian psychologist from Macquarie University, provided more realistic exercise evidence following a four-month training program in financial self-restraint:

Each participant met with the experimenter, individually, at the start, and together they reviewed the participant's bills and spending habits and devised a personal money management plan. Each participant was issued a spending diary and other logs to improve record keeping, both in order to improve adherence to the money self-regulation plan and to enable the researchers to keep track of behavior and performance. Most participants improved substantially in regulating their use of money. Though their incomes did not increase, they spent less and saved more. On average they improved each month and ended up more than quadrupling their savings rate (from 8% to 38% of income).

But what is really amazing is that randomly selected subjects who underwent the money-management training exhibited less ego depletion when they were subsequently tested. Oaten reported that they "got progressively better on laboratory tests of self regulation." Unlike their untrained counterparts, they were able to withstand a thought-suppression task (don't think about white elephants) and still resiliently show self-control (ignore a comedy video and instead follow an electronic shell game).

The exercise research is still in its infancy. But maybe humans can learn to walk and chew gum at the same time. In chapter 1, we learned that four-year-olds who could wait a bit longer before caving in and eating a marshmallow ended up with higher average SAT scores and a better ability to concentrate. But it just may be possible to train kids (and older humans) to have more willpower. The initial empiricism on ego depletion suggests that people like our president should think twice before taking on too many commitments at once. The exercise results, however, suggest that going out of your way to exercise self-restraint—even on arbitrary low-stakes tasks—may make it easier for you to follow through when something important comes along. Under this reading, training himself to be healthier with regard to smoking might help Obama persevere in fighting for health-care reform.

The exercise idea would have been quite congenial to Franklin's own way of thinking. His long-term, systematic efforts at moral perfection smack very much of training. As to his results, Franklin concluded,

On the whole, tho' I never arrived at the perfection I had been so ambitious of obtaining, but fell far short of it, yet I was, by the endeavour, a better and a happier man than I otherwise should have been if I had not attempted it; as those who aim at perfect writing by imitating the engraved copies, tho' they never reach the wish'd-for excellence of those copies, their hand is mended by the endeavor, and is tolerable while it continues fair and legible.

DON'T EVEN THINK ABOUT IT

The idea of ego depletion has even been applied to discrimination. The basis for this set of experiments is that people may have to use up some of their self-control resources when they interact with people of a different race. For example, Baumeister found that white subjects experienced a fall in blood glucose levels after they were asked to speak with a black experimenter and state their opinions about racial profiling and affirmative action. But the decrease in glucose was not uniform. Subjects who independently had obtained a high score on an instrument designed to measure "internal motivation to respond without prejudice" had no substantial drop in glucose. Baumeister's hypothesis, which was confirmed in the experiment, is that people with an independent commitment to respond without prejudice would not have to expend as many resources in a difficult conversation with a person of another race, but that people with a lower commitment would have to use up more resources. It's not just the difficulty of speaking about race; it's the difficulty of speaking about race to someone of a different race. These same low-commitment types, chillingly, showed no glucose drop when they spoke about affirmative action or racial profiling with members of their own race.

Baumeister has shown that ego depletion can also arise in trying to suppress the urge to use stereotypes when describing "Sammy." A different set of subjects were shown a picture of a young man, named Sammy, who was described as gay. The subjects were told to write an essay describing what Sammy does in a typical day, but they were instructed "not to make any mention of stereotypes or any activities they believed homosexuals tend to do." Baumeister also had the subjects independently answer questions from the HATH (Heterosexual Attitudes Toward Homosexuals) instrument to again try to capture their "internal motivation to respond without prejudice." Blind evaluation found that the subjects were generally successful in following the assignment and showed no statistical difference in essay compliance between subjects with higher or lower motivations to respond without prejudice. But the low-motivation subjects had a harder time solving anagrams after the essay writing than they had before. People with a lower internal motivation to respond without prejudice seem to have used more resources to suppress the urge to include stereotypes and had a harder time following through in solving the seemingly unrelated anagram task, which required concentration and a different kind of stick-to-it-iveness.

Baumeister, in separate testing, has been able to show once again that a few spoonfuls of sugar delivered in sweetened lemonade or that a couple of weeks of training on a self-control task can help those with low motivation avoid ego depletion. But to my mind the more interesting possibility suggested by these motivation results is that it might be possible to use commitment contracts to manipulate the level of effort expended in self-regulating choice. The exercise results suggest that, like Benjamin Franklin, people can increase their capacity for self-regulation. But the motivation results suggest that people might be able to reduce their demand for self-regulation. People with an intrinsic commitment to respond without prejudice didn't have to use up their limited capacity of willpower in deciding whether to suppress the urge to stereotype.

People who have internally committed not to steal don't have to fret about whether the coast is clear so that they can pilfer from the cash register. Theft is just not an option, so they don't have to think about it. It might be that commitment contracts can reduce ego depletion by taking some questions off the table. That's the core difference between commitments and mere incentives. As we learned in chapter 2, incentives merely try to guide future choice, but there is still a choice to be made. Commitment contracts, by threatening a carrot too good to refuse or a stick too bad to accept, take a future choice off the table. The threat, for example, of forfeiting a substantial sum of money to a cause you detest may eliminate any question that you will bite your fingernails again and hence not force you to expend limited self-regulatory capacity in the struggle to decide.

At the moment, this is pure speculation. I'm hoping that the incredibly prolific Professor Baumeister will test whether it is true. But the bigger lesson is that before taking on another commitment, it is worthwhile to consider whether it will aid or undermine your efforts to follow through on other aspects of your plan for the good life. The ego exercise story counsels toward even seemingly arbitrary impositions of self-control to help your will stay in training. But the dizzying examples of ego depletion suggest that, as Franklin learned long ago, it might be better not to take on too much, too soon.

GET HAPPY

You can't commit to being happy. Sure, I could write a contract literally committing to be 20 percent happier next month (or forfeit so much money to charity). But no one does this. The pursuit of happiness is one of the ultimate things we strive for, but credible commitments are limited to externally verifiable behaviors and events. Because happiness is internal, it flies below the radar of commitments. The best we can do is commit to do things that we think will make us happy. But here is where the frailty of human judgment again rears its ugly head.

We routinely mispredict what will affect our happiness. In some ways, this is good. Humans are remarkably resilient creatures. We overestimate the sadness that will ensue after bad things happen to us. In *You're Stronger Than You Think*, Peter Ubel details the science of emotional resilience. If you are in a car accident later today and have to have your leg amputated or become paraplegic, there is a good chance you will find new ways to be about as happy as you were before the accident— even if you have to give up your job.

Years ago, a law professor friend of mine told me with total seriousness that if he could get a job teaching at Yale "he'd be happy every day of his life." This is nonsense. When people win the lottery, their average reports of happiness increase, but the increases are remarkably small and transient. Psychologists call this the "hedonic treadmill"—we tend to adapt to changed circumstances and emotionally, as Philip Brickman and Donald Campbell put it, "stay pretty much in the same place."

I can honestly say that commitment contracts have changed my life. My weight roller coaster has evened out. I'm exercising more. But I can't say that I've experienced a profound increase in my overall happiness. I was pretty happy before the commitments—abundantly blessed with a loving family and satisfying work. And I remain pretty happy (despite continuing bouts with insecurity and frustration), even after reengineering several dimensions of my routine. I feel a certain satisfaction and accomplishment, but happiness—like humor—is for me an ephemeral feeling that vanishes if overanalyzed. In retrospect, the commitment that may have brought me the most pleasure is not the one that led to losing weight or exercising more, but the commitment to read novels.

If you take on a serious commitment thinking that it is going to dramatically increase your happiness, you are likely to be disappointed—even if you succeed.

The new science of commitment that I've tried to describe in this book offers a tool to help you reach your goals. Part of this science is learning when commitment contracts are likely to be counterproductive. Alfie Kohn, the author of *Punished by Rewards,* appropriately warns that we should be wary of the "sugar-coated control" of incentive and commitment arrangements. Commitment contracts are likely to backfire if they destabilize social norms and intrinsic motivation for action.

But even when commitments succeed, there is the deeper question of whether the underlying goal is itself salutary. Instead of maximizing happiness with the help of commitment contracts, we might be better off "satisficing."

> Success is getting what you want.
> Happiness is wanting what you get.
>
> ANONYMOUS

Barry Schwartz, who chided the Yankees for insisting on incentives in Joe Torre's contract, published a disturbing article suggesting that goal seeking itself might be inimical to the pursuit of happiness. Barry followed students at eleven different colleges over the course of their search for postgraduation employment. At the beginning of the school year, he gave seniors a questionnaire asking whether they agreed or disagreed with statements like "When I am in the car listening to the radio, I often check other stations to see if something better is playing, even if I am relatively satisfied with what I'm listening to" and "When shopping, I have a hard time finding clothes that I really love." He used their answers to help divide them into two groups: "maximizers" and "satisficers." Maximizers are more likely to undertake an exhaustive search of all possibilities, while satisficers are more likely to search only until they have found a "good enough" option.

At the end of the year, Barry went back to see how the students did in their job search. As you might imagine, the maximizers ended up snagging jobs that paid more than the jobs offered to satisficers. A lot more. After controlling for more than a dozen other characteristics (including things like GPA and college major), Barry found that the maximizers' average starting salaries were $7,430 higher than those of seniors who were independently coded as being satisficers—more than 20 percent higher.

But the real surprise came when Barry asked the seniors how they felt. "Despite their relative success, maximizers [were] less satisfied with the outcomes of their job search, and more pessimistic, stressed, tired, anxious, worried, overwhelmed, and depressed throughout the process," he reported. Maximizers did better financially but felt worse. Maximizers' pursuit of "the elusive best" systematically increased their regret. Their high-paying jobs didn't live up to the maximizers' even higher expectations.

Committing to secure the very best may be a recipe for disaster. Professor Schwartz counsels instead that we try to train ourselves to be satisficers—to reduce our consideration of the grass that might be marginally greener on the other side of the fence. Even here, commitment contracts might help. One can imagine commitments to satisfice that might, for example, force us to sample at most three stations before settling in to listen to a song. Some of the seniors in Schwartz's study might even have been better off if they had committed to forfeit any salary beyond a certain amount—so that they would have less regret if they failed to achieve the very highest salary.

But while it is easy to conjure new arenas where commitment devices might improve happiness, it might be even wiser to remember that we humans also have an abiding desire to keep some spheres of our lives beyond the reach of incentives and commitments. From time to time, we like to take literal and figurative vacations from our normal world of obligation. We choose to go on cruises or to all-inclusive resorts in part because they offer a place where our choices are unpriced. We are attracted to the idea of "what happens in Vegas stays in Vegas" in part because it suggests a taste of decisions without consequence. Although we admire Benjamin Franklin, we should think twice before emulating his attempt to scrutinize and constrain every aspect of his behavior. In a chapter that speaks at such length about ego depletion, we would do well to remember that our personalities are composed not just of egos (and superegos) but also of ids that want to be free.

A Commitment Store

I first met Dean Karlan on a hot summer day in 2006 when we decided to have lunch in the beautiful English garden next to Yale's School of Management. Dean was wearing a faded T-shirt and shorts, and he looked more like a camp counselor than a college professor. My first impression was that he was the cuddliest of academics, with a warm smile and a laugh that came easily (even in response to his own jokes).

Dean also has the soft-spoken enthusiasm of someone who knows when he has an idea worth pursuing. It's a trait that has served him well. A couple of years after graduating from college with a degree in foreign affairs, Dean followed his passion and worked for three years on village banking with FINCA International, one of the leaders in alleviating poverty with microfinance solutions in developing countries. He then decided that he needed some more tools and returned to school, simultaneously earning master's degrees in business and public policy from the University of Chicago. But here's where Dean's independent streak really kicked in. While he was studying at Chicago, he got increasingly interested in behavioral economics and field experiments—and instead of cashing in on his MBA or returning to the nonprofit sector, Dean decided to get a PhD in economics. He wanted to learn how to test what policies really worked. So at age thirty-one, when most people are years out of school, Dean launched himself into a new degree program.

When I had lunch with him in 2006, Dean had transformed himself into one of the hottest young economists on the planet. Before getting tenure he won the highest honor given by the federal government to young researchers, the Presidential Early Career Award for Scientists and Engineers. This prize (which comes with $400,000) is rarely given to social scientists; that Dean was honored with it is a tribute to the impact his field experiments are having on real-world policy in the developing world.

But at this lunch Dean wasn't talking about collaborating on an academic study. Instead, he pitched the idea that we sink thousands of dollars into a new business start-up. There, among a profusion of roses, he described offering an Internet service that would allow anyone to craft a real commitment contract. Instead of the Money Store, which provides loans, Dean's vision was to create a commitment store to provide one-stop shopping for people who want to commit to real change.

The idea of commitments was anything but new. But Dean saw that the relatively friction-free world of the Internet could be used to make it a lot easier for people to make commitments. The commitment store could provide standardized terms and a reputation for enforcement. From the very beginning, however, Dean wanted to get out of the way and let the users decide the direction of the site. He predicted that the biggest success might ultimately be for types of contracts we had never imagined. He saw the site as a means of connecting commitment partners who might not even have heard of each other in advance.

The more Dean spoke about a commitment store, the more I found myself caught up in the idea. I felt an immediate connection with Dean. Here was a kindred spirit who liked going out and crunching numbers on things that impact people's lives.

Dean's belief in the power of commitments comes from personal experience. "When I was in the third year of grad school at MIT," Dean said, "I noticed that I weighed more than I ever had before." One of his close friends had the same problem. Dean recalled, "We were dismayed about how our studies had gotten the better of our bodies." He and his friend took decisive action. They entered into a contract where Dean committed to go from 208 to 170 pounds and his friend committed to lose the same amount, to go from 218 to 180 pounds. We know from chapter 5 that Dean's goal to lose more than 18 percent of his body mass was aggressive, and the contract also had an aggressive speed. He and his friend had to lose three pounds every two weeks. But what was really aggressive about the contract was its stakes. Dean and his friend put at risk half of their annual income. They were in graduate school, so this wasn't Donald Trump money, but failure would drastically change their already spare lifestyles—plus, neither wanted to pay thousands of dollars to another economist and become the punch line of a bad joke.

"But the funny thing is," Karlan remembered, "at first, neither of us lost weight. We hung around near each other and realized that neither of us was doing anything." They ended up repeatedly renegotiating and extending the deadline. So version 2.0 of their commitment contract added a clause providing that any attempt to renegotiate would trigger an immediate forfeiture of that person's stakes. "It worked," Dean said. "We both lost the weight, weighing in at 180 and 170 in January of 2002."

Their success was short-lived. "We said we'd enter into a maintenance arrangement," Dean noted, "but we didn't." A bit like Ryan Benson from season one of *The Biggest Loser*, they both quickly gained about twenty pounds back before they confessed to each other what had happened. What's worse, they stopped keeping in touch. "We went a long time," Dean said, "without talking to each other, something we both realized was a true sign that the other was gaining weight." After several months, they finally put a new contract in place, with stakes still set at half their annual income. But by this time they both had jobs, and the potential amount at risk was eye-opening. Once again, the contract worked wonders. They both lost the weight and kept it off—with an exception that proves the rule-like nature of the commitment. In the middle of the commitment, Dean's friend bounced up a bit—enough so that he owed Dean $15,000. Dean was happy to take it. "If I didn't," Dean explained, "no future contracts would ever work." Many thought Dean's friend would be mad or upset, but he was pleased that Dean took the money. He knew this would kick him into shape.

The idea behind stickK in many ways started with Dean's struggle to find a mechanism that would actually work. To be effective, you need to have a credible threat. You need to believe that if you fail, you'll actually have to forfeit your stake. Dean quickly learned that one of the biggest problems with private commitment contracts is that it's often hard to find a partner who will follow through on the punishment.

A TREE GROWS IN MINDANAO

Dean's faith in the power of commitments also grows out of his own field experiments. In 2002, Dean partnered with the Green Bank of Caraga, a rural bank in Mindanao, in the Philippines, to test the impact of savings commitments. He explained, "We wanted to test out a very simple idea: Does making savings illiquid...help people save more? So we took about 1,800 former and current clients of a bank and we randomly took about half of them and offered them a commitment savings account." The new account, called a SEED (Save, Earn, Enjoy Deposits) account restricted the clients' access to their own deposits until they met their savings goals. At the time of setting up the account, the client could choose to block his or her own withdrawals until after a specific date or until after a specific amount had been saved.

Dean quickly learned that there was a substantial market demand for the commitment service. "We found," Dean said, "that about 28 percent of the people who were offered the account took up the offer. It was a very simple idea that immediately struck a chord with a lot of people." Of the people who opted for the new account, about 70 percent chose a specific date as their ending goal (instead of a specific amount to be saved). The relative popularity of the SEED account is especially striking because it paid no higher interest than an ordinary account. "Traditionally," Dean explained, "if we are going to get a bank account that is more illiquid, like a certificate of deposit, we are going to actually receive a higher interest rate to compensate us for tying up our funds." But in Mindanao, Dean found people who were willing to restrict their future freedom for no extra compensation.

This high take-up rate is particularly interesting for someone who is thinking about starting a commitment store, because it suggests that a substantial group of consumers want to tie their own hands. This shouldn't come as too much of a surprise. Christmas club savings accounts and store layaway plans have long provided consumers with illiquidity commitments. In *A Tree Grows in Brooklyn,* the stalwart mother is forever adding coins to a tin-can bank nailed to the floor of a closet in the family's tenement flat. The nails that fasten the can to the floor serve as a commitment (to keep the father from squandering the family's savings on drink), because the money can be retrieved only with difficulty, by prying the can from the floor. People who opened SEED accounts weren't given a tin can and nails, but they were offered (for about a dollar) a *ganansiya* box. "It is similar to a piggy bank," Dean said. "It has a small opening to deposit money and a lock to prevent the client from opening it. In our setup, only the bank, and not the client, had a key to open the lock. Thus, in order to make a deposit, clients need to bring the box to the bank periodically." More than 80 percent of the SEED account holders opted for this additional savings aid.

Dean went further and found that the consumer demand came from just the people Thaler would have predicted—those with time-inconsistent preferences. Before making any offers, his research team asked all 1,800 bank clients a monetary version of Thaler's apple questions:

(1) Would you prefer to receive 200 pesos guaranteed today or 300 pesos guaranteed in one month?

(2) Would you prefer to receive 200 pesos guaranteed in six months or 300 pesos guaranteed in seven months?

More than a quarter of the respondents made the patient choice for the second question but were impatient in answering the first question, where the possibility of an immediate payoff was more tempting. What's more, Dean found that women who exhibited time-inconsistent preferences were much more likely to sign up for the SEED account. "People who know they are going to be tempted," Dean said, "actually prefer to have fewer choices in the future." For someone who was kicking around the possibility of a commitment store, the SEED study can be read as a kind of market research. Dean not only learned that there can be substantial demand for commitment products, he also figured out more about whom to target.

The truly remarkable findings came when Dean looked at the average savings of the treatment and the control groups. Just being offered the commitment device caused an increase in average savings. After six months, the average savings of the group offered SEED accounts was 47 percent higher than the savings of the control group, and after a year, this difference ballooned to an 82 percent increase. It's notable that these percentages are the average savings increases of everyone who was offered a SEED account, including the bank clients who turned down the offer. When Dean looked just at those who actually opened SEED accounts, he found that their savings increased more than threefold (1,715 pesos) relative to the control group. While an extra 1,700 pesos is not much by American standards (it's about $34), in this area of the Philippines, where a doctor's visit costs about $3 and a one-month supply of rice for a family of five costs $20, it is enough to make a difference.

The SEED account is a classic disabling device. Dean called his article on the experiment "Tying Odysseus to the Mast" because, like Odysseus, account holders who disabled withdrawals until a specific date could do nothing but wait for the self-imposed constraint to be lifted. But Dean followed up the SEED experiment with a test of a commitment device that is much closer to the kind of commitment that would eventually become a central part of stickK.com.

Working again with Green Bank, Dean developed a savings account to help people quit smoking. Dean said, "We offered a very simple savings account to clients that said, 'Look, put money into the savings account and in six months we are going to give you a urine test. If you are not smoke-free from this urine test, then we are going to actually take your money out of your savings account and donate it off to a local orphanage. And you will lose your money.'" The bank called this strange new animal the CARES account (an acronym for "Committed Action to Reduce and End Smoking").

To really tease out whether the CARES account actually worked, Dean again set up a randomized experiment. Green Bank representatives approached smokers in public places and asked them to take a short survey. All the smokers were given an informational pamphlet on the dangers of smoking and a tip sheet on how to quit. But at random, only half the smokers were offered the opportunity to set up a CARES account.

Again, Dean found that a surprisingly high proportion of those offered the CARES account accepted the offer: 11 percent of smokers decided to put money into an illiquid zero-interest account where they could get their funds back only if they passed a nicotine and cotinine test. In some ways, that's even more impressive than the 28 percent take-up rate on the SEED offers. As Dean emphasizes, "The CARES account was a totally foreign concept. It isn't just a slight variation on an existing consumer product that is out there. I mean, this really had to be explained to people and to have one in nine people agree to do something that is totally new to the market ... I think any consumer market company would be thrilled to know that their initial intervention had an 11 percent take-up rate ... among their target audience."

Dean thinks that a desire to control your future self is again driving the demand for the commitment device. "When someone is sophisticated enough to realize that they have a problem, they will want to increase the future price of cigarettes," Dean said. "But you can't do that—we can't go to cigarette companies voluntarily and say, 'Please charge more.' You can't go to your local store and say, 'When I come to you in a week and ask you to buy a cigarette, please charge me twice as much.' The store won't do that." The CARES account helps by increasing "the price of vice."

To see whether CARES worked, Dean went back six months later and tested whether there was nicotine in the urine of three different groups: those who used the CARES account, those who were offered the account but declined, and those in the control group, who were never offered the account. Dean found that people who signed up for the account were about 30 percent more likely to quit smoking than either the control group or those who turned down the CARES offer. Dean said, "Thirty percent is a huge number for a smoking-cessation project." Putting at risk, on average, 550 pesos—just $11—was enough to increase the success rate from 7 percent to about 37 percent. "That's a higher success rate than is generally seen among those who try to quit smoking using nicotine patches," said Steve Levitt, the economist author of *Freakonomics*. Putting more money at risk was correlated with a higher probability of success. The CARES account holders who succeeded averaged more than $20 at risk, while those who failed averaged less than $6 at risk.

For those who worry that the impact of this kind of extrinsic motivation would be short-lived, Karlan went back at the end of the year and paid both the treatment and the control groups to participate in a second round of surprise urine tests. The good news is that even six months after the commitment ended, the CARES account holders were 30 percent more likely to have clean tests. The CARES "treatment" was not a panacea—more than 60 percent of the account holders still failed to present a clean urine sample. But in a world where the success rate of smokers who use the nicotine patch or gum often hovers below 10 percent,

Dean's results from the small island of Mindanao suggest that a bank savings account can be a useful aid to help people kick the habit.

MARKET TESTS

When Dean described the initial results of the smoking study, back during our first lunch, I remember feeling a literal tingle in the small of my back. Barry Nalebuff (he of the teach-a-Yale-class-in-a-Speedo challenge) was also interested. Barry and Dean had once stayed up until one A.M. brainstorming about commitments. And while Barry and I weren't sure that a commitment store could produce a viable revenue stream, we also figured that anything that could make progress on the very difficult problems of weight loss and smoking cessation had more than a chance of paying for itself.

When it came time for Barry and me to write our next *Forbes* column, we decided to tackle the hard question of whether a commitment store could both be attractive to users and make money. Our column, "Skin in the Game," began:

> Imagine a Commitment Store that offered a financial incentive to lose weight. You promise to lose a pound a week for the next 20 weeks and then keep the 20 pounds off for the remainder of the year. You back up the promise with a $1,000 weight-loss bond. Weigh-ins are biweekly. Every time you meet your goal, you get back $60. Over the course of the year you could earn back $1,560. Of course, each time you miss, that costs you $60. Here's a diet system that literally pays you to lose the weight.

It sounds crazy to offer 56 percent annual interest. But Barry and I knew that very, very few diets succeed. Even if half the dieters succeeded under this new regime, the store would still turn a healthy profit as the recipient of the forfeited stakes. Putting aside the interest it could earn on deposited funds (and the overhead costs), it would be taking in $2,000 from a pair of dieters and paying out $1,560 to one of them.

I was unusually proud of what we had written. But when I read the final version that was prepublished on the *Forbes* website a day before the paper version hit the newsstands, my pride turned to deep chagrin. Somehow, in the process of writing and editing the piece, all references to Dean and his research had been left on the cutting-room floor. Barry and I, like dozens of other economists, had thought long and hard about commitment contracts, and Barry had independently thought from time to time about providing commitment services. But I still couldn't believe that the column had been published without giving Dean some credit for the idea of a commitment store.

I remember with a kind of visceral dread running across campus to Dean's office to tell him about my mistake. Dean reacted with preternatural calm and after a few seconds turned to his computer and simply said, "We should grab the domain name." Unfortunately, we found that the URL www.commitmentstore.com had already been purchased—just a few hours earlier! Someone had seen an online version of the column and gone out and bought the domain name. I immediately told Dean that I would take responsibility for buying it back. Thankfully, the purchaser just wanted to indicate his love of the idea and transferred ownership to Dean and me for free. Oddly, the first market test—which Dean's idea passed—was caused by my stupidity. Dean had every right to be furious at me. I expressly gave him the option of jettisoning me from the start-up. But in a strange way, the incident (including my abject apology and his forgiveness) drew us closer together.

Our next challenge was to find a CEO. Dean and I could put up money and throw some time into the company, but we already had day jobs as professors. If the commitment store was going to be a reality, we'd have to convince someone to put his or her heart and soul into its creation. Luckily, we found Jordan Goldberg. Jordan had graduated from Yale College in 2006 and Yale gave him a free ride for half of his MBA tuition as one of just six "silver scholars." During the spring of 2007, as Jordan was finishing his first year of business school, we sat down with him and started brainstorming. Before we knew it, he had signed on and taken an open-ended leave from business school. If stickK continues the way it's going, he may never go back. We formed an LLC, opened a bank account, and took the plunge, throwing thousands of dollars of our own money into the idea.

Even though Dean and I had had some anxious moments over losing the URL www.commitmentstore.com, Jordan soon had us convinced that we needed a name that was a lot shorter. The idea of a store for commitments might have been the metaphor that led Dean to originally think up the business, but people don't like typing in all those letters. I remember brainstorming with Jordan in my law school office coming up with dozens of alternative names—and nervously checking whether the domain names were still available. I tried riffing on different variants of "carrots and sticks." I particularly liked Lady Macbeth's admonition to "screw your courage to the sticking place." Our website could be the sticking place. Sadly, the site name for the traditional spelling of "stick" was already being used (by the manufacturer of a bodyless guitar called "the stick"). But it was Jordan who suggested that we use the name stickK. Conjuring an alternative spelling is very Web 2.0. And "stickK" is a wonderfully short double entendre. We're not only trying to help people stick to their goals; we offer contracts that use sticks instead of carrots. (In fact, the extra K is also the traditional legal denotation for a contract—so stickK is offering stick contracts.)

The real fun was sitting down with Dean and Jordan and working late into the night, hammering out the nitty-gritty details of how to structure the site and the contracts themselves. Our bedrock principle was to put our users in charge. "Really, the heart of the idea is that it is totally voluntary," Dean explained. "We wanted a site that allows people to align their incentives with what they actually say they really want to do. . . . It is a way of making your vices more expensive and your virtues cheaper." You select the goal and the type of accountability you want. So you not only decide when you want to quit smoking or how much weight you want to lose (within reason), but you also choose whether you want to put money at risk. Or, alternatively, you might give us the email addresses of supporters. "These are people," Dean explained, "who are simply informed of your success or failure. For a lot of people, the social pressure is perfectly sufficient." That's what's at stake if you fail to meet your commitment.

Part of choosing the type of accountability is picking the person who gets to decide whether you kept your commitment: your referee. You can choose to be your own referee, with an "on your honor" contract, or you can designate anyone else in the world to be your ref just by giving us his or her email address. At first, we worried about what to do with contracts where the designated person refused to act as a referee. But we handled this problem just by *presuming* that referee silence means that the client is keeping the commitment. Setting the default meaning of referee silence to be success means that I can create a binding year-long exercise contract with stickK designating Warren Buffett as my referee. Even if the Oracle of Omaha doesn't initially respond to stickK's referee request, I'm still contractually bound to exercise, and there's a chance that at some point during the contract Mr. Buffett will register and report on how I'm doing. Because of the default, we can create binding contracts without waiting for the referees to agree to perform their task.

In designing a site where people put real money at risk, we understood that it was crucial to figure out clear-cut conditions for forfeiture. We'd drown in disputations if we had to make thousands of Solomonic judgments about whether a particular user had kept his or her commitment. We needed authoritative indicators of failure. In the end, we decided to build in three forfeiture triggers. stickK would forfeit your stakes if: (1) you failed to report within forty-eight hours of your report's due date; (2) you reported that you failed to stick to your commitment; or (3) for refereed contracts, your referee reported that you failed to stick to your commitment. The third trigger is the most problematic. We don't second-guess the refs. If a user says she succeeded, but her referee reports that she failed, we forfeit the stake. We make this clear in the terms and conditions:

> Client agrees that reports from the Referee shall be binding. . . . Client acknowledges that stickK . . . has no duty to verify the accuracy of the Referee's reports. . . . [I]t is a basic assumption of the parties in making this contract that Client will bear the risk that the Referee reports to stickK might be inaccurate.

The good news is that so far, at least, clients have accepted the judgments of their referees. "We definitely see," Jordan said, "instances where the referee overrides the user's report, where the user reports success and the referee says, 'No, you failed!' But we have not had one person come and dispute a referee's report of failure."

Probably the most important question was deciding who gets the forfeiture money. My *Forbes* column imagined that all forfeited funds would go to the commitment store itself—and we envisioned using some of that forfeited money to pay successful clients more than they had originally put at risk. But this was the one part of our pitch where people pushed back. They didn't like the idea that we would profit from their failures. And we worried that people wouldn't entrust their stakes to a start-up company that was taking the other side of the bet. So we decided not to take forfeited stakes, with the exception of a transaction processing fee, and instead to allow the users to designate the charity (or anti-charity) or individual of their choosing. In order to make money, we would focus on ad revenues and a premium service that allows businesses and organizations to offer customized contracts and virtually any combination of carrots and sticks to motivate employees and customers.

IF YOU BUILD IT, THEY WILL COME

There were many late nights in the fall of 2007 while we "specked out" the site's pages. After considering a variety of options, including in-house programmers, we contracted with a Montreal-based programming firm, Summit Tech, to bring our specs to life. After spending Thanksgiving with his family, Jordan packed a few clothes and started driving to Canada. "I decided I wasn't coming back until the site is launched," Jordan remembered. "I literally called the developers on the way and said, 'I am on my way to Montreal.' I didn't even have a place to stay. That first night, actually, one of the developers let me crash at his place." In the end, Dean and I would also make trips to Montreal to help keep pushing. We really wanted to launch in time for New Year's resolutions.

It might not be the best of omens for a commitment site to miss its own first big commitment. But the site features were not quite stable enough by New Year's, and so it was not until January 14, 2008, that we opened the virtual doors to the public. We initially offered standardized contracts to lose weight, quit smoking, or exercise. Plus, we let users custom-design their own contracts with "I commit to" followed by a blank text box they could fill in as they saw fit.

I was more than a little scared that the opening would be like a tree falling unheard in the wilderness. Maybe people won't trust us with their money or their credit card numbers, I thought; maybe people in the United States just don't want these kinds of commitments. What's worse, we had no advertising budget. Dean and I wrote a couple of op-eds and I blogged about the launch, but mostly we were hoping that the idea of letting people make real commitments backed by money would catch people's imagination.

It definitely made good copy for newspapers and magazines. Since our launch, stories everywhere from the *Economist* and NPR's *All Things Considered* to *O, the Oprah Magazine* and *Esquire* have spread the word. The site experienced huge spikes in registrations after TechCrunch, the most popular tech blog, touted the service. When Yahoo included our link in an article on its home page, listing stickK as one of five new ways to save money, 2,000 people signed up in just two hours. In our first three months, users trusted us to hold more than $250,000 of their money.

We've also been able to sell the concept to investors. We sold 10 percent of the company in December 2007, even before launching, and were able to sell another 10 percent during the worst of the market meltdown in November 2008. As an ivory-tower academic, I sometimes have a hard time believing that what started with a few thousand dollars and a lot of sweat equity has turned into a business worth millions of dollars.

To be clear, stickK's usage numbers are nothing like Facebook's gargantuan statistics. But as I write this, in August 2009, still several months shy of our second birthday, stickK has more than 40,000 registered users from more than 130 countries and we're growing at an annual rate of more than 125 percent.

Some of our users are supporters or referees. And some just registered without having made contracts of their own. But we've created a site where people have entered into more than 25,000 contracts. The average contract lasts about three months. Our users are also opting for all the mechanisms of accountability. More than a quarter of our contracts have a referee. And slightly less than a quarter have designated supporters who will learn of the user's success or failure. More than 10,000 of these contracts are backed by financial stakes; so far, stickK users have put more than $3 million at risk. When they do put money at risk, they are averaging a nontrivial $275 per contract.

Word of the site traveled very, very quickly among both economists and contract scholars, who are fascinated with the idea of bringing to life a business founded upon an academic idea. At Harvard, stickK inspired an entire law school exam; it quotes extensively from stickK's terms and conditions and asks students, in true Socratic fashion, how courts would respond to all manner of hypothetical events. But not all academics were originally smitten with the idea. University of Chicago professor Eric Posner questioned whether our contracts were legally enforceable:

> There is no consideration for your promise to pay money to StickK if you fail to lose weight—no quid pro quo, which is legally required for an enforceable contract. StickK does not give you anything in return for your promise to lose weight; nor do you give anything to StickK in return for its promise to pay the charity if you fail to lose weight. A rare zero-sided contract, there is no quid and no quo.

These are, in an academic sense, fighting words. In my day job, I teach and have written extensively about contract law. I am the coauthor of one of the leading casebooks from which contract law has been taught to generations of first-year students. It would be a blunder of monumental proportions if I'd drafted the "rare zero-sided contract" in which there is neither a quid nor a quo—especially since I teach my students that any lawyer worth his or her salt can usually conjure consideration with a little thought. Contrary to Posner's claim, it is easy to find both quids and quos in the stickK terms and conditions. Most centrally, the clients promise to meet their commitment and stickK promises to email friends and supporters if the clients fail in their commitment. A lot more could be said, but stickK is happily contracting with thousands of clients in its second year, taking in millions of dollars of stakes and valid credit card promises, and has not encountered a single problem of legal enforceability.

Other economists have written that the site will never work because an evil doppelgänger business, unstickK.com, will spring into being to render stickK commitments ineffective. For example, Australian economist Joshua Gans wrote:

> Put simply, if the idea of StickK.com is to allow people to commit to a plan and penalty, unStickK.com destroys it as it will be always worthwhile for you to purchase a counter-contract from them after you have "committed" to one from StickK.com. Moreover, the very existence of an unStickK option means that

consumers will not find it worthwhile to purchase a contract from StickK.com. So neither StickK.com nor its anti-counterpart will end up making money.

But this is crazy talk. The premium for such insurance would be outrageous. The fancy economics term for this is "adverse selection"—because the people who are willing to pay for unstickK insurance are precisely the ones who are looking to break their initial promises. Think about our New Zealand friend James Hurman, who auctioned his smoking addiction. How much would you demand to take over his obligation to pay $1,000 per cigarette smoked for the rest of his life? Suffice it to say that no unstickK product has ever been offered. At the end of the day, stickK, like many Internet start-ups that simply give away their services, might not survive. But it won't be because of nonenforceability or anti-commitment insurance.

Instead of dreaming up "business-killing" ideas, the George Mason University economist Alex Tabarrok suggested a perverse but complimentary business, one in which he simply serves as an unsavory destination for forfeited stakes. When another economist offered to designate Alex as the recipient of $100 placed on a stickK weight-loss contract, Alex responded:

> I understand the difficulty of losing weight thus I want you to know that if we receive the $100 we will *not* send it to India, we will *not* give the money to cancer research, we will *not* give the money to any cute animals instead we will use your money to squash the poor, to fight against universal health care, and to gas up our Hummer. Moreover, we will do this while drinking fine wine, smoking cigars, eating foie gras and laughing uproariously. There that ought to help. If there are other left-wingers out there who would like more motivation to accomplish their life goals then do know that [I am] here to help.

Alex's "new business idea" is mostly tongue-in-cheek. But it suggests another type of motivation. When you set up a financial commitment at stickK, you can choose the "friend or foe" option—where you can designate an individual or even a for-profit company to be the recipient of any forfeited stakes. In the Hurman auction, people compete for this honor by offering you cash up front. Alex perversely sees that there might alternatively be a kind of repugnance competition, where potential recipients compete for designation by promising to do outrageous things with the money.

Although Alex is targeting the "left-wing" commitment market, repugnance competition need not know a particular political persuasion. For example, one of the organizers of the blog Weight upon the Lord, which seeks to support and encourage "Christian men in their journey to maintain good physical and spiritual health," reported on his own stickK contract to lose twenty-four pounds in twelve weeks:

> That's just two pounds a week, which should be entirely doable. And if it isn't? Well, the William Jefferson Clinton Presidential Library will be getting some of my money! . . . I really don't want to give any money to honor someone with whom I disagree on many, many topics, so hopefully this will be the extra boost I need to stick to my program. Each time I'm tempted to eat something I shouldn't, or to skip a workout, I'll just picture good ol' Bubba munching on a cheeseburger and fries—that oughta do it.

Overall, the anti-charity feature has been popular, being used in more than 40 percent of the contracts backed by financial stakes. And while images of our forty-second president help motivate some, it is our forty-third president who has been the real uniter among stickK users. The George W. Bush Presidential Library has far and away pulled in the largest number of anti-charity dollars.

At the end of the day, what's most important is whether the commitment contracts help people stick to their goals. We'd view it as a huge setback if people forfeited the vast majority of their money. But the first results are heartening: only about 20 percent of the money put at risk is being forfeited; stickK users have reported losing almost sixty thousand pounds; and our reported success rate for users who are trying to quit smoking and back up their commitment with cash is well over 60 percent.

Let me quickly emphasize that all these are *reported* successes. There is every chance that when push comes to shove, some people lie and falsely report a success to avoid losing money. Still, the supercynical should take note that 19 percent of our users in "on your honor" contracts still choose to forfeit their own money. At least some people cannot bring themselves to lie.

These success statistics will not be truly credible until we have independent verification that the reported successes were actual successes. To this end, stickK is collaborating with academics to collect the data to do more rigorous analysis. But in the interim, there are hopeful signs. For example, for one-shot contracts where users put money at stake, the success rate is nearly identical whether or not the contract was refereed by a third party. Without a referee (the "on your honor" contracts), there is every opportunity for people to lie. The 82 percent success rate on these contracts must be viewed with some skepticism. With referees, there is less opportunity to avoid forfeiture by falsehood, because a user's referee can blow the whistle on any false report of success. But here, we find that the success rate remains at a remarkably constant 81 percent. Of course, even referees can be co-opted.

On New Year's Day 2009, Justin Noble, a public health student in Toronto, designed a commitment contract to go to church once a week. Justin had successfully used stickK before to get to work on time. To make sure his commitment to go to church stuck, he backed it up with all three accountability devices that the site has to offer. He put down $600, and he signed up his girlfriend as both a supporter and a referee to make sure he followed through. The contract seemed to work like a charm. Every week, Justin faithfully reported to us that he had succeeded. But when I called Justin to ask him about his experience with stickK, he confessed that he had "cheated" in some of his reports to stickK. Even though the commitment at least indirectly implicates a higher, all-seeing referee, Justin (with his girlfriend's complicity) was willing to send in false reports. Indeed, Justin thinks that setting such a stiff financial penalty made it too hard for his future self to report failure.

Justin's story is a cautionary tale not just about the success statistics; it's another example of what can happen when you put too much pressure on commitment devices. Justin entered into too many commitments. Once he and his girlfriend realized that they could bring themselves to lie, the commitments lost their force. But Justin's story has a bit of a silver lining. After taking a break from using stickK, he found that he could come home again, and that the site is again helping him stick to his goals.

Justin's story also underscores the importance of designating multiple supporters for your commitment. Remember how the hard part for Andy Mayer wasn't creating the stickK commitment but emailing all his friends that he had done it? It's easier to conspire with your girlfriend than with an entire congregation. Justin's church commitment would probably have ended very differently if he had added his pastor as one of his supporters. I used to think that having a referee was the most important complement to a financial stake. But the stories I've uncovered of referees letting their friends slide make me now think that having multiple supporters may be as important. When it comes to commitments, there's safety in numbers.

Stepping back, I see that the first results suggest that all three forms of accountability matter. Backing up your promise with financial stakes has the biggest impact. But even without money at risk, having supporters and a referee helps. On one-shot contracts, the success rate without supporters is 45 percent; with one supporter, it is 52 percent; but with two or more supporters, the reported success rate rises to 60 percent. And people who have referees are more likely to report success than those who are on their honor: 62 percent versus 37 percent. In contrast, users who opt for contracts without money at risk, without supporters, and without a referee, have less than a 25 percent chance of success (however, many of these users may just be testing out our system).

POLYMORPHOUSLY PERVERSE

One of the most pleasurable exercises in writing this book was reading through more than eight thousand of the custom-designed contracts where stickK users filled in their own commitment devices. Each contract tells a story of individual struggles to change lives for the better. The contracts reveal the immense diversity in the human condition. stickK has attracted widely divergent types of people. Some people come to the site to make grand and amorphous contracts to self-actualize and improve their personal lives. Others come with narrow and precise work goals: for example, to make one hundred sales calls per week.

We've already seen that users are motivated by diametrically opposed anti-charities. Some can't stand giving to the NRA; others don't want their money going to the gun-control organization Stop Handgun

Violence. But there is also a yin and yang to the commitment goals themselves. Some people commit to losing weight, while others want to bulk up. Some people commit to playing fewer video or board games, while others commit to reaching new "techno trigger" levels on their DS. One of the more gendered dichotomies concerns dating. Women have made close to a dozen commitments not to call former boyfriends, while men have made close to a dozen commitments to ask women out on dates.

In May 2008, we reviewed a commitment from a user (who has asked to remain anonymous) putting $100 at risk in an eight-week, refereed commitment not to manipulate the glabella. To be honest, this commitment sent us to the dictionary to learn that the glabella is "the space between the eyebrows and above the nose." This commitment is just one of hundreds of attempts to curtail bad habits and compulsions. Users commit to stop biting their nails, cracking their joints, and picking their skin. Haley Cohen, a college sophomore, committed to stop saying "like" inappropriately. She failed miserably and wrote about her experience for her college magazine. She likes the site, but thinks it will work only if you independently want to stick to your goal. "There is no way that an extrinsic service or motivator," she said, "is going to be able to force you to do something. You have to have some sort of internal, intrinsic motivation to do it." Then again, she put only $9.99 at risk because "$10 just seemed too scary to send the George Bush Library." That's not much extrinsic motivation. It's almost like she planned to fail (and write the article). She might be right, but it would have been a fairer test of the proposition if she had put something more meaningful at risk—say, $5,000. In fact, the success rates were substantially lower on contracts where people put less money at risk. For one-shot contracts like Haley's, the success rate where less than $20 was put at risk was 68 percent, but with more than $100 at risk, it rose to 81 percent.

It's easy to scoff at some of these commitments to break bad habits. In the big picture, they are small potatoes. But I've also read about heartrending attempts of people using commitments to try to stop binge eating or blacking out from drinking too much alcohol. Several clients have used commitments, backed by money, to try to rein in a gambling problem. One user even committed to stop cheating on his wife.

There are also hundreds upon hundreds of contracts to take action. Scott Bierko of Yorktown Heights, New York, put $50 at stake and committed to writing four songs in four weeks. Scott is a children's singer-songwriter who performs with his wife as "Beth and Scott" at arts and education assemblies in upstate New York. But his contract was to write songs for grown-ups. "It's something I've always wanted to do," he told me, "and I really felt that stickK was a good way to hold my feet to the fire—to actually fulfill a commitment that I had some inhibitions about." So Scott and another songwriter entered into contracts to be creative, to write music. "We were both able to complete our goal," Scott said. "It was a fantastic experience." Something as nerdy as a commitment contract could spur his muse.

Other people have committed to write dissertations, to do their taxes, to send thank-you notes, and to floss. Harvard behavioral economist David Laibson was able to finally clean his office. People commit to learn languages—French, Spanish, and C++—and they commit to practice things like guitar, saxophone, and yoga. Nick Knol committed to write an iPhone app for Allstate. Ari Dworkin-Cangor, probably our youngest user, at age seven (even though our terms limit use to people thirteen and up!), successfully used stickK, with the help of her psychologist mother, to start using "a fork instead of hands at dinner." As foreshadowed in previous chapters, some people commit to inputs (committing to study more), others to achieve results (committing to raise their GPA). Some of the anti-procrastination commitments concern the small annoyances in life that we put off, but behind these simple contracts we also see people committing to file for child support, to apply for career-changing jobs, and to pay off crippling bills.

Lauren Harrison, a recent college graduate who'd just started working as a management consultant in Chicago, committed to calling her grandmother at least once a week. "It brings her great pleasure and doesn't take a lot of my time," Lauren told me. This simple contract with small anti-charity stakes and automated reminders has rekindled her relationship with her nana. "Sometimes the phone calls were just a few minutes just for me to say hello," Lauren reported, "And sometimes they were much longer. It was definitely the first time in years where I had called my grandmother that regularly. Before that I would call her maybe once a month. Sometimes once every two months. And usually at the prodding of my mother, or I would get a nagging voice-mail message from my grandmother herself saying, you know, 'I haven't heard from you in a while. I want to make sure that you are still alive.' So, it felt good to do that little thing and she was thrilled and getting those reminders and having that financial incentive was hugely instrumental in doing it."

Lauren is one of several people who've committed to calling their grandmothers or mothers. Her story is one of hundreds of commitments that are directed primarily at improving the lives of others. There is no easy or fast distinction between self-directed and other-directed commitments. When we help others, we often help ourselves, and vice versa. Calling regularly brought Lauren great happiness. In retrospect, she says, "I can't believe for how many years this was something that I had done nothing about."

Lauren's story also shows the viral quality of these commitments. Lauren made her boyfriend the referee for her contract. He saw first-hand how well the contract was working, and soon he asked Lauren if she would be his referee on a commitment to reduce his drinking. The contract helped him cut down, but he still failed from time to time. You might think that this would put Lauren in an awkward situation. "Actually, him signing up for it," she told me, "and formally choosing me as a referee gave me the freedom to nag. I was just no longer a nagging girlfriend. Instead, I was basically fulfilling my end of the deal to serve as his referee. So actually it created some space for me to offer my input about that in a way that I couldn't before. Or I didn't want to before." Becoming his referee reduced the tension in the relationship. Even their willingness to report honestly about his failures signaled something to each other about their integrity. "I felt at first that being his referee might increase contention, but it turned out to be totally fine because he wanted me to do that job."

Not every stickK story is a success, and with a little prodding, I found out that even some of the reported successes were in fact failures. But stories like Lauren's make me think that stickK on net is a force for good in the world. It has helped real people make progress on real problems.

These stories only scratch the surface of the breadth of the human condition that can be found in stickK contracts. Dean's favorite involves someone who committed to speaking more slowly to foreigners in New York City. One can also find commitments to dunk a basketball, to grow a ponytail, or, as *Seinfeld* put it, to be the master of your domain. People in the throes of change curse frequently and conjure wonderfully colorful descriptions of what they are fighting. You can find "drybruary" commitments (not to drink in February) and "McVommit" commitments, which speak for themselves.

Like Benedick in *Much Ado About Nothing* (who enjoined his companions to "pick out mine eyes with a ballad-maker's pen and hang me up at the door of a brothel-house" if he ever were to look pale with love), one user has created a commitment to stay single. But wonderfully and perversely, on the same site you can also find commitments to treat one's spouse more kindly and to "try for a baby."

THE FUTURE

Moving ahead, I see increasing signs of commitment contracts being used by businesses to help their customers and employees effect change. Your electric company can help you commit to reduce your bill by 10 percent, or your employer might help you commit to carpool to work. Your local car dealer might help you save for that new car. Your boss might even help you remember to call your grandmother. But increasingly, we may also find companies going further and demanding commitments that are primarily for the company's good.

Some life insurance companies are already offering lower prices to people who aren't overweight at the time of signing up—but they might offer even lower prices to people who enter into credible commitments to stay skinny or smoke-free. Your boss may "help" you commit to meeting a project deadline. Recently, stickK launched a premium version of the site that will allow employers to use carrots as well as sticks to change behavior. In a world where the unhealthy lifestyles of individual employees can add thousands of dollars every year to an employer's health costs, incentivized wellness programs are almost certain to be on the rise. Matching programs where employees put some of their own money on the line but can be rewarded by their employer for success are especially likely to provide a powerful behavioral bang for the buck.

The American Cancer Society is using stickK's new rewards platform to help its donors lead healthier lives. Instead of hitting up your friends to donate to the ACS for every mile you cover in a walkathon, you can create a stickK contract where your friends automatically give to the

ACS if you quit smoking for the next twelve months. And you can back up your commitment by promising, in the same contract, to pay the ACS a bunch of your own money if you fail to kick the habit.

Beyond wellness, businesses can design their own campaigns to turn stickK into a performance-management tool to make sure that deliverables really get delivered. In January 2010, the good people at Staples launched a national "StickK to It" Commitment Challenge to help their customers achieve their career and business goals. The challenge makes it easy for small businesses and individuals to get organized and increase their productivity by committing to take specific and concrete steps, such as:

to update my résumé by the end of the week,

to print using recycled paper and recycle my ink and toner cartridges,

to set up regular invoicing and collection processes,

to scan my receipts, enter expenses into QuickBooks, and file all bills,

to compile all my contacts' information from business cards into one location,

to create a monthly budget forecast for the calendar, or

to set up a secure wireless network to make sharing printers easier.

Staples is enticing people to participate by offering them "EasyPoints" for creating and following through on their commitments. The Easy-Points can be redeemed online for anything from a donation to the Boys and Girls Clubs to a raffle ticket for a chance to win a laptop. My kids' favorite item is the famous Staples "Easy Button."

The campaign is a natural for Staples. Sometimes mere procrastination keeps Staples's customers from reorganizing their lives or their business processes. The campaign's carrots help customers do what they really have been wanting to do all along (and helps Staples sell a few products to help them do it). Like other aspects of the Staples brand, the campaign is trying to improve the customers' experience with annoying or painful professional tasks. With a few clicks at stickK, the reaction can now be "That was easy."

As we've seen, commitment devices are not just ways of changing our own lives; they are ways, for good or ill, that we can demand credible change from others. In the future, I expect to see employers use commitment sticks to incentivize superior job performance. Managers might, for example, challenge their sales force to make a certain number of calls—or else. Some of these demands should be resisted. But the potential of commitments as a screening device to change behavior is too powerful and effective to leave completely off the table. And fair notice: stickK, with its corporate portals, hopes to become an important player in this space.

We also hope to provide our users with better information about the likely outcome of making different kinds of commitments. For example, just before entering into a one-shot contract, you might find it useful to know that your probability of success will rise by 8 percent if you add an additional supporter. Ideally, we'd like to give weight-loss users the entire bell curve of probable reactions—including, especially, the chance that they might end up weighing more. We're not there yet. But the database we are amassing, limited as it is by self-reports of success and failure, is a powerful engine for finding out what works.

This book is full of tips on how to better tailor incentives, commitments, and even anti-incentive devices to better achieve your goals. But the truth is that behavioral economics and psychology is still a young science. My advice on when to use carrots versus sticks, and when to use commitments at all, has been at times impressionistic. It is appropriate to end on a note of modesty and caution. The key is to continue testing. We want to find out whether telling future people about past results can make them more sophisticated. We want to find out when we could have altered the terms of the contracts to change some of our failures to successes.

But a hidden part of me is not modest or cautious. In 2008, Harald zur Hausen won the Nobel Prize in medicine for discovering how the human papillomavirus can cause cervical cancer. To my mind, it was a prize he richly deserved. But if we can develop commitments that routinely produce even 50 percent success rates at losing weight or quitting smoking, stickK could be responsible for saving many, many more lives. That would be a powerful medicine. In such a world, people like Richard Thaler and even Dean Karlan, who dreamed of a commitment store, would be worthy contenders as well.

Acknowledgments

ASSISTED LIVING

My sainted father always resisted the idea of assisted living. He worried about the loss of freedom. Not me. At the tender age of fifty, I've already actively sought living assistance in many areas of my life. Sarah Politz helps nag me when I stop reading novels. Sonia Cintron keeps our domestic life in order. Merry Maids descends on our house regularly to attack the dirt. The whole idea behind stickK.com is a kind of assisted living—to assist people to be their best selves.

My own efforts to write this book have been assisted by many others. My agents, Lynn Chu and Glenn Hartley, continue to speak bluntly about what half-baked ideas don't work. They will not serve up a proposal before its time. At Yale Law School, my longtime assistant, Marge Camera, transcribed innumerable interview tapes on subjects lapsing further and further away from law. And a gaggle of truly excellent research assistants, led by Anna Arkin-Gallagher, painstakingly proofread and gap-filled the beast you have now finished. I also benefited from the gentle prodding of the students in my Incentives and Commitments seminar, who forced me to organize my thoughts about this subject, but who also went into the field to collect data for more than a dozen projects.

However, the biggest assist came from my editor, John Flicker. John learned a bit about stickK while editing the afterword to my book *Super Crunchers*. It was John who pushed me to write this book and to write it now. I tend to agree with more than 80 percent of his suggestions—an extraordinary rate in my experience with editors, and a rate that makes me think twice before deciding that a suggestion falls in the 20 percent. But as important as the substance of his suggestions is that John has been there for me when I needed him most.

Finally, it's not easy living with a man obsessed with incentives and commitments. My poor kids, Anna and Henry. As children, they have suffered through incentive schemes where they had to earn their first TV (by sitting through a semester of corporate finance), their first dog (by coauthoring a journal article), and their second dog (by taking the AP history exam)—all without the protection of a university human subjects committee or the city's child services department. If Alfie Kohn is right that children are *Punished by Rewards,* I will have surely retarded their intrinsic motivation to learn.

And through it all there has been Jennifer Gerarda Brown, my coauthor and beloved spouse. She has had to put up with my worst "pay me now or pay me later" impulses. Countless times I've interrupted her by asking, "How do I make this transition?" or "Can I talk you through the next section?" I would be lost without her.

In high school, I inscribed a quotation from the Spanish philosopher José Ortega y Gasset on my backpack: "Life is a desperate struggle to be in fact that which we are in design." The older I get, the more overblown the quotation seems. My assisted life is at most one of muted desperation. But still, after all these years, the quotation has some resonance for me. My hope in writing this book and in helping to create stickK is to help you be in fact what you are in design.

Notes

INTRODUCTION: SNEEZE

xiv **Curt Schilling's weight-loss contract:** Gordon Edes, "Sources: Schilling Out Until at Least All-Star Break," *Boston Globe,* Feb. 8, 2008, p. E1.

CHAPTER 1: THALER'S APPLES

3 **Thaler's letter:** Richard Thaler, "Some Empirical Evidence on Dynamic Inconsistency," *Economics Letters* 8 (1981): 201. A more standard starting point would be Thaler's 1980 article "Toward a Positive Theory of Consumer Choice" in the very first volume of the *Journal of Economic Behavior and Organization.* There are many other starting points, including: Richard H. Thaler and H. M. Shefrin, "An Economic Theory of Self-Control," *The Journal of Political Economy* 89 (1981): 1392, and D. Kahneman and A. Tversky, "Prospect Theory: An Analysis of Decision Under Risk," *Econometrica* 47 (1979): 263. For a history of the field, see Colin F. Camerer, George Loewenstein, and Matthew Rabin, eds., *Advances in Behavioral Economics* (2003). More information on Thaler and hyperbolic discounting can be found in David Laibson, "Golden Eggs and Hyperbolic Discounting," *Quarterly Journal of Economics* 112 (1997): 443, and Roger Lowenstein, "Exuberance Is Rational," *New York Times Magazine* (Feb. 11, 2001). I thank Dean Karlan for pointing me to the apple example.

4 **Save More Tomorrow:** Richard H. Thaler and Shlomo Benartzi, "Save More Tomorrow: Using Behavioral Economics to Increase Employee Saving," *Journal of Political Economy* 112 (2004): S 164.

6 **April Fool's:** Ted O'Donoghue and Matthew Rabin, "Doing It Now or Later," *American Economic Review* 89.1 (1999): 103.

7 **Increasing impatience as time horizons shrink:** Partha Dasgupta and Eric Maskin, "Uncertainty and Hyperbolic Discounting," *American Economic Review* 95 (2005): 1290.

7 **Pigeon experiments:** George Ainslie and R. J. Herrnstein, "Preference Reversal and Delayed Reinforcement," *Animal Learning and Behavior* 9 (1981): 476; George Ainslie, "Impulse Control in Pigeons," *Journal of the Experimental Analysis of Behavior* 21 (1974): 485; George Ainslie, "Specious Reward: A Behavioral Theory of Impulsiveness and Impulsive

Control," *Psychological Bulletin* 82 (1975): 463; George Ainslie, "Beyond Microeconomics: Conflict Among Interests in a Multiple Self as a Determinant of Value," in *The Multiple Self,* ed. Jon Elster (1986); George Ainslie, *Picoeconomics* (1992); Robert H. Strotz, "Myopia and Inconsistency in Dynamic Utility Maximization," *Review of Economic Studies* 23 (1956): 165.

10   **Humans impatient for cash gift:** George Ainslie and V. Haendel, "The Motives of the Will," in *Etiology Aspects of Alcohol and Drug Abuse,* ed. E. Gottheil, K. Druley, T. Skodola, and H. Waxman (1993). But see Mara Airoldi, Shane Frederick, and Daniel Read, "Longitudinal Tests for Inconsistent Planning Due to Hyperbolic Discounting," http://papers .ssrn.com/sol3/papers.cfm?abstract_id=1287233 (2008), which criticizes aspects of Ainslie and Haendel's design. Other sources also explore the time inconsistency: Serdar Sayman and Ayse Onculer, "An Investigation of Time-Inconsistency," working paper, http://papers.ssrn.com/sol3/ papers.cfm?abstract_id=1013210 (2007); and R. H. Strotz, "Myopia and Inconsistency in Dynamic Utility Maximization," *Review of Economic Studies* 23 (1956): 165.

10   **Pregnant women and anesthesia:** J. J. Christensen-Szalanski, "Discount Functions and the Measurement of Patients' Values: Women's Decisions During Childbirth," *Medical Decision Making* (1984): 47; Thomas C. Schelling, *Strategies of Commitment and Other Essays* (2006), 63.

12   **Of mice and men:** This phrase is often paraphrased in English as "The best-laid plans of mice and men / Go often wrong." http://en.wikipedia .org/wiki/To_a_Mouse.

12   **"part of the hard-wiring":** R. Frank, *Passion Within Reason: The Strategic Role of the Emotions* (1988).

12   **"vertebrates devalue delayed events":** George Ainslie, "Précis of *Breakdown of Will*," *Behavioral and Brain Sciences* 28 (2005): 635, 649; George Ainslie, "Specious Reward: A Behavioral Theory of Impulsiveness and Impulsive Control," *Psychological Bulletin* 82 (1975): 463; R. Frank, *Passion Within Reason: The Strategic Role of the Emotions* (1988).

12   **Four-year-olds and marshmallows:** Mischel and his colleagues also tested children's willingness to delay gratification with regard to one or two pretzels and one or two colored poker chips. Walter Mischel, Yuichi Shoda, and Monica L. Rodriguez, "Delay of Gratification in Children," *Science* 244 (1989): 933; Yuichi Shoda, Walter Mischel, and Philip K. Peake, "Predicting Adolescent Cognitive and Social Competence from Preschool Delay of Gratification: Identifying Diagnostic Conditions," *Developmental Psychology* 26 (1990): 978; and Walter Mischel, Yuichi Shoda, and Philip K. Peake, "The Nature of Adolescent Competencies Predicted by Preschool Delay of Gratification," *Journal of Personality and Social Psychology* 54 (1988): 687.

12   **Ringing the bell:** Yuichi Shoda, Walter Mischel, and Philip K. Peake,

"Predicting Adolescent Cognitive and Social Competence from Preschool Delay of Gratification: Identifying Diagnostic Conditions," *Developmental Psychology* 26 (1990): 978, 980.

13 **"exhibit[ed] self-control":** Yuichi Shoda, Walter Mischel, and Philip K. Peake, "Predicting Adolescent Cognitive and Social Competence from Preschool Delay of Gratification: Identifying Diagnostic Conditions," *Developmental Psychology* 26 (1990): 978, 983 table 2.

13 **"The causal links":** Walter Mischel, Yuichi Shoda, and Monica L. Rodriguez, "Delay of Gratification in Children," *Science* 244 (1989): 933, 936.

14 **"exhibit a tendency":** Ted O'Donoghue and Matthew Rabin, "Optimal Sin Taxes," *Journal of Public Economics* 90 (2006): 1825, 1828.

14 **Richard Thaler, of course, understands:** George Loewenstein and Richard H. Thaler, "Anomalies: Intertemporal Choice," *Journal of Economic Perspectives* 3 (1989): 181; and Richard H. Thaler and Cass R. Sunstein, *Nudge: Improving Decisions About Health, Wealth, and Happiness* (2008).

15 **"you must bind":** Robert Strotz opened his 1956 article on dynamic inconsistency with the same quotation from Homer's *Odyssey* chapter 12. Robert H. Strotz, "Myopia and Inconsistency in Dynamic Utility Maximization," *Review of Economic Studies* 23 (1956): 165. See also Hammond Innes, *The Conquistadors* (1969), 70–73; Andrés de Tapia, "The Chronicle," in *The Conquistadors: First-Person Accounts of the Conquest of Mexico,* ed. and trans. Patricia de Fuentes (1963), 25–26; and Robin Lane Fox, ed., *The Long March: Xenophon and the Ten Thousand* (2004). For more on the writings of Xenophon, see John Dillery, *Xenophon and the History of His Times* (1995). I learned of Cortés and Xenophon in Avinash Dixit and Barry Nalebuff, *The Art of Strategy* 16 (2008) and from Thomas C. Schelling, *Strategies of Commitment and Other Essays* (2006), 1, 72 (discussing Xenophon and various methods of incapacitation).

16 **"To a Mouse":** Robert Burns, "To a Mouse," www.worldburnsclub.com/poems/translations/554.htm.

17 **"tie their beaks":** George Ainslie, "Impulse Control in Pigeons," *Journal of the Experimental Analysis of Behavior* 21 (1974): 485, 487.

17 **"The tendency to seek":** George Ainslie, "Impulse Control in Pigeons," *Journal of the Experimental Analysis of Behavior* 21 (1974): 485.

18 **Rabin's website:** Matt Rabin's website can be found at: www.econ.berkeley.edu/~rabin/index.html. The capital of Montana is Helena.

19 **Preproperate:** Ted O'Donoghue and Matthew Rabin, "Doing It Now or Later," *American Economic Review* 89 (1999): 103, 111 n. 16. Ted told me that after they had made up this word, they found a few ancient uses of it in the *OED.*

20 **The good as enemy of the great:** *"Le mieux est l'ennemi du bien,"* Voltaire, *La Bégueule* (1772).

CHAPTER 2: INCENTIVES VERSUS COMMITMENTS

24 **"the carrot and the stick":** Steve Shavell first pointed this out to me. The earliest citation of the expression's use to describe alternate rewards and punishments was recorded by the *Oxford English Dictionary* as from an article in the December 11, 1948, issue of the *Economist*. The article in the *Economist* referred to "The material shrinkage of rewards and the lightening of penalties, the whittling away of stick and carrot." "Carrot, n." *Oxford English Dictionary* (1989).

24 **The benefits of incentives:** In fact, the internalization tool described in chapter 4 of *Why Not?* ("Why Don't You Feel My Pain?") is a core tool for promoting creativity. Barry J. Nalebuff and Ian Ayres, *Why Not? How to Use Everyday Ingenuity to Solve Problems Big and Small* (2003).

24 **The dangers of dog poop:** J. A. Marron and Charles L. Senn, "Dog Feces: A Public Health Concern and Environmental Problem," *Journal of Environmental Health* 37 (1974): 239. The Centers for Disease Control and Prevention website lists the numerous diseases that can be carried by dogs; see www.cdc.gov/healthypets/animals/dogs.htm.

25 **Dr. Bar-On's program:** Avida Landau, "City Uses DNA to Fight Dog Poop," Reuters, Sept. 16, 2008, http://uk.mobile.reuters.com/mobile/ m/AnyArticle/p.rdt?URL=http://uk.reuters.com/article/lifestyleMolt/idU KLG37942520080916; and Rebecca Skloot, "The Dog-Poop DNA Bank," *New York Times Magazine*, Dec. 12, 2008.

25 **The London congestion charge:** "Who Pays What," Transport for London, www.tfl.gov.uk/roadusers/congestioncharging/6741.aspx (last visited on May 30, 2009); and "Congestion Charging in London," *BBC News*, http://news.bbc.co.uk/1/shared/spl/hi/uk/03/congestion_charge/ exemptions_guide/html/what.stm (last visited on May 30, 2009).

26 **Quote from Maimonides:** Isadore Twersky, *A Maimonides Reader* (1972): 404–7.

26 **Paying kids to do well in school:** Jennifer Medina, "For 'A' Students in Some Brooklyn Schools, a Cellphone and 130 Free Minutes," *New York Times*, Feb. 28, 2009; and press release, NYC Department of Education, "Chancellor Klein Launches 'Million' Motivation Campaign," Feb. 27, 2008, http://schools.nyc.gov/Offices/mediarelations/NewsandSpeeches/ 2007–2008/20080227_million.htm.

27 **D.C.'s Capital Gains program:** "Capital Gains Program: Frequently Asked Questions," District of Columbia Public Schools, www.k12.dc.us/ capital-gains/faqs.htm (last visited May 30, 2009).

27 **"disruptive, profane":** "Pay-to-Behave Program Debuts in D.C. Schools," NPR.com, Oct. 21, 2008, www.npr.org/templates/story/story.php ?storyId=95949912.

27 **paying students for grades in Israel:** Joshua Angrist and Victor Lavy, "The Effects of High Stakes High School Achievement Awards: Evidence

from a Group-Randomized Trial," *American Economic Review* 99 (2009): 1384.

27 **Colbert interviews with Fryer:** *Colbert Report* (Comedy Central television broadcast, Dec. 1, 2008), available at www.colbertnation.com/the-colbert-report-videos/164944/december-01-2008/roland-fryer (last visited on Oct. 9, 2009).

28 **Bounty on rat pelts:** Fred Gottheil, "UNRWA and Moral Hazard," *Journal of Middle Eastern Studies* 42 (May 2006) 409–21; and Robert K. Merton, "The Unanticipated Consequences of Purposive Social Action," *American Sociological Review* 1 (1936): 894; and Ian Ayres and Peter Cramton, "Pursuing Deficit Reduction Through Diversity: How Affirmative Action at the FCC Increased Auction Competition," *Stanford Law Review* 48 (1996): 761.

28 **Schelling on commitments:** Thomas C. Schelling, *The Strategies of Commitment and Other Essays* (2006).

29 **Committing not to drink with Antabuse:** George Sereny, Vidya Sharma, James Holt, and Enoch Gordis, "Mandatory Supervised Antabuse Therapy in an Outpatient Alcoholism Program: A Pilot Study," *Alcoholism: Clinical and Experimental Research* 10 (1986): 290.

29 **"alli oops":** "Alli Oops! I Just Pooped Myself," Pharma Marketing Blog, http://pharmamkting.blogspot.com/2007/02/alli-oops-i-just-pooped-myself.html; and Stephen J. Dubner and Steven D. Levitt, "Freakonomics: The Stomach Surgery Conundrum," *New York Times Magazine*, Nov. 18, 2007.

30 **Alli's success:** "Common Questions About the Alli Weight Loss Program—How Effective Is the Alli Program?," www.myalli.com/whatisalli/commonquestions.aspx#AL2. A clinical study conducted over a three-year period concluded, "Two-year treatment with orlistat plus diet significantly promotes weight loss, lessens weight regain, and improves some obesity-related disease risk factors." Participants in the study lost on average 4–8 percent of their body weight, depending on the course that they followed. See also "Weight Control and Risk Factor Reduction in Obese Subjects Treated for Two Years with Orlistat," *JAMA* 281 (1999): 235.

31 **Kosher phones:** Steven Erlanger, "A Modern Marketplace for Israel's Ultra-Orthodox," *New York Times*, Nov. 2, 2007; "Markets in Everything," *Marginal Revolution*, www.marginalrevolution.com/marginalrevolution/2007/11/markets-in-ever.html; and MIRS Communication, www.mirs.co.il/.

31 **The kosher phone lawsuit:** Complaint, *Yeshiva Yagdil Torah v. Sprint Solution Inc.*, Civil Action No. 06-CV-13726 (Dec. 4, 2006), available at www.media.nymag.com/docs/06/12/KOSHERPHONE.pdf.

33 **Zappos' incentive to quit:** Keith McFarland, "Why Zappos Offers New Hires $2,000 to Quit," *BusinessWeek*, Sept. 16, 2008; and William C. Taylor, "Why Zappos Pays New Employees to Quit—and You Should Too,"

Practically Radical, May 19, 2008, http://discussionleader.hbsp.com/taylor/2008/05/wy_zappos_pays_new_employees.html. Despite its revolutionary incentive tactics, Zappos has recently undergone a round of layoffs. See C. V. Harquail, "If Stephen Colbert Were the CEO of Zappos: Explaining a Layoff to Your Employees," *Authentic Organizations,* Nov. 13, 2008, http://authenticorganizations.com/harquail/2008/11/13/if-stephen-colbert-were-the-ceo-of-zappos-explaining-a-layoff-to-your-employees/.

33   **Zappos' customer service miracle:** Stephen J. Dubner, "Customer Service Heaven," *New York Times* Freakonomics blog, May 17, 2007, http://freakonomics.blogs.nytimes.com/2007/05/17/customer-service-heaven/; see also Steven D. Levitt, "Amazing Customer Service," *New York Times* Freakonomics blog, Sept. 29, 2008, http://freakonomics.blogs.nytimes.com/2008/09/29/amazing-customer-service/.

35   **The I-15 story:** Lior Strahilevitz, "How Changes in Property Regimes Influence Social Norms: Commodifying California's Carpool Lanes," *Indiana Law Journal* 75 (2000): 1231. Strahilevitz relied on data from Janusz Supernak et al., "I-15 Congestion Pricing Project Monitoring and Evaluation Services, Phase I Overall Report" (1999).

37   **The rich uncle:** *Hamer v. Sidway,* 124 N.Y. 538 (1891).

40   **Liability rules and property rules:** Guido Calabresi and A. Douglas Melamed, "Property Rules, Liability Rules and Inalienability: One View of the Cathedral," *Harvard Law Review* 85 (1972): 1089.

40   **"smoker's entrance fee":** John Whitehead, "Friday Beer Post," *Environmental Economics,* Nov. 25, 2008, www.env-econ.net/2008/11/friday-beer-p-2.html; and "Fine, Thank You," *Economist.com,* Dec. 4, 2008, www.economist.com/blogs/freeexchange/2008/12/fine_thank_you.cfm.

41   **On managers' ethical obligations:** While the authors do not discount managers' desire to conform to ethical standards, they point out the fallacy of claiming that a corporation would have moral obligations. Frank H. Easterbrook and Daniel R. Fischell, "Antitrust Suits by Targets of Tender Offers," *Michigan Law Journal* 80 (1982): 1155, 1171, 1177 n.57.

43   **Lump sum payments:** Ted O'Donoghue and Matthew Rabin, "Studying Optimal Paternalism, Illustrated by a Model of Sin Taxes," *American Economic Review* 93 (2003): 186; see also Jay Bhattacharya and Darius Lakdawalla, "Time-Inconsistency and Welfare," National Bureau of Economic Research, Working Paper No. 10345 (Mar. 2004), www.nber.org/papers/w10345; and Lee Anne Fennell, "Slices and Lumps," University of Chicago Law and Economics, Olin Working Paper (Mar. 1, 2008), http://papers.ssrn.com/sol3/papers.cfm?abstract_id=1106421.

43   **Einstein's view of insanity:** "Insanity: doing the same thing over and over again and expecting different results," Quotations Page, www.quotationspage.com/quote/26032.html.

CHAPTER 3: LOSSES LOOM LARGE

46 **Risking money to lose weight:** Robert W. Jeffery, Wendy M. Gerber, Barbara S. Rosenthal, and Ruth A. Lindquist, "Monetary Contracts in Weight Control: Effectiveness of Group and Individual Contracts of Varying Size," *Journal of Consulting and Clinical Psychology* 51 (1983): 242. See also Robert W. Jeffery, Paul D. Thompson, and Rena R. Wing, "Effects on Weight Reduction of Strong Monetary Contracts for Calorie Restriction or Weight Loss," *Behaviour Research and Therapy* 16 (1978): 363, which analyzes a similar study conducted with severely obese adults, where those who entered into commitment contracts to lose weight lost more weight than those who did not.

47 **Typical weight loss in studies:** See, for example, Kishore M. Gadde, Deborah M. Franciscy, H. Ryan Wagner II, and Ranga R. Krishnan, "Zonisamide for Weight Loss in Obese Adults: A Randomized Control Trial," *JAMA* 289 (2003): 1820; N. D. Luscombe, P. M. Clifton, M. Noakes, E. Farnsworth, and G. Wittert, "Effect of a High-Protein, Energy-restricted Diet on Weight Loss and Energy Expenditure After Weight Stabilization in Hyperinsulinemic Subjects," *International Journal of Obesity* 27 (2003): 582.

47 **"Forbidden toy" studies:** E. Aronson and J. M. Carlsmith, "Effect of the Severity of Threat on the Devaluation of Forbidden Behavior, *Journal of Abnormal and Social Psychology* 66 (1963): 584; Elliot Aronson, "The Power of Self-Persuasion," *American Psychologist* (November 1999): 875, 876; E. Scott Geller, "The Art of Self-Persuasion" (2001), www.safety performance.com/pdf/Articles/2001/TheArtofSelf-Persuasion.pdf; J. L. Freedman, "Long-Term Behavioral Effects of Cognitive Dissonance," *Journal of Experimental Social Psychology* 1 (1965): 145; and Mark R. Lepper, "Dissonance, Self-Perception, and Honesty in Children," *Journal of Personality and Social Psychology* 25 (1973): 65.

48 **Wine choice:** Eric van Dijk and Daan van Knippenberg, "Trading Wine: On the Endowment Effect, Loss Aversion, and the Comparability of Consumer Goods," *Journal of Economic Psychology* 19 (1998): 485. The wine experiment is an update of the pioneering endowment test: Jack L. Knetsch, "The Endowment Effect and Evidence of Nonreversible Indifference Curves," *American Economic Review* 79 (1989): 1277.

49 **Monkey Loss Aversions:** Keith Chen et al. "How Basic Are Behavioral Biases? Evidence from Capuchin Monkey Trading Behavior," *Journal of Political Economy* 114 (2006): 517.

50 **Thaler and Kahneman's experiments:** Daniel Kahneman, Jack L. Knetsch and Richard H. Thaler, Experimental Tests of the Endowment Effect and the Coase Theorem," *Journal of Political Economy* 98 (1990): 1325. For more on the endowment effect, see Amos Tversky and Daniel Kahneman, "Loss Aversion and Riskless Choice: A Reference-Dependent Model," *Journal of Quarterly Economics* 106 (1991): 1039; Dan Ariely, Joel

Huber, and Klaus Wertenbroch, "When Do Losses Loom Larger Than Gains?" *Journal of Marketing Research* 42 (2005): 134; Daniel Kahneman, Jack L. Knetsch, and Richard H. Thaler, "Anomalies: The Endowment Effect, Loss Aversion, and Status Quo Bias," *Journal of Economic Perspectives* 5 (1991): 193; and Ziv Carmon and Dan Ariely, "Focusing on the Foregone: How Value Can Appear So Different to Buyers and Sellers," *Journal of Consumer Research* 27 (2000): 360. Not all scholars support the endowment effect theory. See, for example, Charles R. Plott and Kathryn Zeiler, "The Willingness to Pay/Willingness to Accept Gap, the 'Endowment Effect,' Subject Misconceptions, and Experimental Procedures for Eliciting Valuations," *American Economic Review* 95 (2005): 530.

50 **The costs of carrots and sticks:** Giuseppe Dari-Mattiacci and Gerrit De Geest, "Carrots, Sticks and the Multiplication Effect," *Journal of Law, Economics, and Organization* 26 (forthcoming 2010).

50 **Higher stakes lower your expected forfeiture:** For those who are mathematically inclined, here is some algebra with which to explore the conditions under which high stakes can lower expected forfeiture. Let the expected forfeiture from a commitment be stated as

$$E(F) = P(S) \times S,$$

where $E(F)$ = the expected forfeiture,
S = the stakes at risk, and
$P(S)$ = probability of forfeiture of stakes given S.

Using the trusty tools of introductory economics, we can minimize the expected forfeiture by finding the stakes that make:

$$dE(F)/dS = P(S) + (dP/dS) \times S = 0 \qquad (1)$$

which can be restated as:

$$P/S = -dP/dS. \qquad (2)$$

If we assume that people will work harder to keep their commitment when the stakes are increased, then dP/dS is likely to be less than zero. Equation (2) implies that the expected forfeiture on a commitment contract will be minimized when:

$$-(dP/dS)/(P/S) = e = 1$$

where e is the elasticity of the probability of forfeiture with respect to the stakes. As long as the probability of forfeiture is elastic, you can decrease your expected forfeiture by increasing your stakes. Or to put it slightly more simply, so long as the percentage reduction in P is greater than the percentage increase in S, you can reduce your expected forfeiture by increasing your stakes. For sufficiently high stakes, there are good reasons to believe that the responsiveness of the probability will become inelastic. But as long as the probability is elastic with regard to stakes, you should raise your stakes. Indeed, given the benefits of changing your behavior, you may well find it to your advantage to increase your stakes above the point that minimizes the expected financial forfeiture. But if you choose to contract, you should at least raise your stakes to the point that minimizes your expected forfeiture.

51 **Preventing premature deaths:** Kevin G. Volpp, M. V. Pauly, George Loewenstein, D. Bangsberg, "P4P4P: An Agenda for Research on Pay for Performance for Patients," *Health Affairs* 28 (2009): 206.

52 **Volpp's weight-loss study:** Kevin G. Volpp, Leslie K. John, Andrea B. Troxel, Laurie Norton, Jennifer Fassbender, and George Loewenstein, "Financial Incentive–Based Approaches for Weight Loss: A Randomized Trial," *JAMA* 300 (2008): 2631.

53 **Volpp on cash-reward programs:** Daniel J. DeNoon, "Dieters Lose Weight When Reward Is Cash," WebMD, Dec. 9, 2008, www.webmd.com/diet/news/20081209/dieters-lose-weight-when-reward-is-cash.

53 **Expected value of one buck:** George Loewenstein, E. U. Weber, C. K. Hsee, N. Welch, "Risk as Feelings," *Psychological Bulletin* 127 (2001): 267; and Todd A. Olmstead, Jody L. Sindelar, and Nancy M. Petry, "Cost-Effectiveness of Prize-Based Incentives for Stimulant Abusers in Outpatient Psychosocial Treatment Programs," *Drug and Alcohol Dependence* 87 (2007): 175.

55 **Volpp on deposit contracts:** Daniel J. DeNoon, "Dieters Lose Weight When Reward Is Cash," WebMD, Dec. 9, 2008, www.webmd.com/diet/news/20081209/dieters-lose-weight-when-reward-is-cash.

57 **Selling a smoking habit:** Smoking Habit for Sale, http://smokinghabitforsale.com/index.html (last visited on June 15, 2009); "Smoking Habit Contract," trademe.co.nz (last visited on June 15, 2009), www.trademe.co.nz/Browse/Listing.aspx?id=146646769; and Robert D. Cooter and Ariel Porat, "Anti-Insurance," *Journal of Legal Studies* 31 (2002): 203.

61 **Variants of the Hurman auction:** Michael Abramowicz independently developed the idea of auction commitments. See Michael Abramowicz and Ian Ayres, "Fair-Bet Commitments: The Law and Economics of Commitment Bonds That Compensate for the Possibility of Forfeiture," working paper (2009), which argues that the government might auction commitment bonds to reduce the deficit.

62 **Weight-loss wagers:** Nicholas Burgery and John Lynhamz, "Betting on Weight Loss . . . and Losing: Personal Gambles as Commitment Mechanisms," working Paper (Jan. 2008), www2.hawaii.edu/~lynham/Research_files/BL_weight01012008.pdf.

62 **The Russian theater and the Asian flu:** Michael Wines, "Chechens Seize Moscow Theater, Taking as Many as 600 Hostages," *New York Times,* Oct. 24, 2002, www.nytimes.com/2002/10/24/world/chechens-seize-moscow-theater-taking-as-many-as-600-hostages.html; Michael Wines, "Hostage Toll in Russia over 100; Nearly All Deaths Linked to Gas," *New York Times,* Oct. 28, 2002; and "Moscow Theater Hostage Crisis," Wikipedia, http://en.wikipedia.org/wiki/Moscow_theater_hostage_crisis (last visited on June 21, 2009). The Asian flu experiments were originated in Amos Tversky and Daniel Kahneman, "The Framing of Decision and the Psychology of Choice," *Science* 211 (1981): 453–58. See also Anton Kühberger et al., "Framing Decisions: Hypothetical and Real," *Organizational Behavior and Human Decision Processes* 89 (2002): 1162–75.

64 **"The perceived risk":** Peter Salovey and Pamela Williams-Piehota, "Field Experiments in Social Psychology," *American Behavioral Scientist* 47 (2004): 488.

67 **Higgins's do-overs:** John M. Roll, Stephen T. Higgins, and Gary J. Badger, "An Experimental Comparison of Three Different Schedules of Reinforcement of Drug Abstinence Using Cigarette Smoking as an Exemplar," *Journal of Applied Behavior Analysis* 29 (1996): 495; Kenneth Silverman et al., "Increasing Opiate Abstinence Through Voucher-Based Reinforcement Therapy," *Drug and Alcohol Dependence* 41 (1996): 157; Stephen T. Higgins et al., "Incentives Improve Outcome in Outpatient Behavioral Treatment of Cocaine Dependence," *Archives of General Psychiatry* 51 (1994): 568; Stephen T. Higgins, Sarah H. Heil, and Jennifer Plebani Lussier, "Clinical Implications of Reinforcement as a Determinant of Substance Use Disorders," *Annual Review of Psychology* 55 (2004): 431; and Barry J. Nalebuff and Ian Ayres, *Why Not? How to Use Everyday Ingenuity to Solve Problems Big and Small* (2003): 21.

68 **Kirby's cocaine study:** Kimberly C. Kirby, Douglas B. Marlowe, David S. Festinger, R. J. Lamb, and Jerome J. Platt, "Schedule of Voucher Delivery Influences Initiation of Cocaine Abstinence," *Journal of Consulting and Clinical Psychology* 66 (1998): 761.

68 **Noncash incentives:** Scott A. Jeffrey, "Justifiability and the Motivational Power of Tangible Non-Cash Incentives," *Human Performance* 22 (2009): 143; Ran Kivetz and Itamar Simonson, "Self-Control for the Righteous: Toward a Theory of Precommitment to Indulgence," *Journal of Consumer Research* 29 (2002): 199; and Paul Nolan, "Why Cash Incentives Fail," *SalesforceXP Magazine,* Sept.–Oct. 2005, www.salesforcexp.com/edit/200509/cover.php.

70 **Ray Romano's golf game:** Rebecca Louie, "5 Minutes with . . . Ray Romano," *New York Daily News,* Apr. 9, 2006.

70 **Bail-bond collateral:** Ian Ayres & Joel Waldfogel, "A Market Test for Race Discrimination in Bail Setting," *Stanford Law Review* 46 (1994): 987.

71 **Finkelstein's incentives study:** Eric A. Finkelstein, Laura A. Linnan,

Deborah F. Tate, and Ben E. Birken, "A Pilot Study Testing the Effect of Different Levels of Financial Incentives on Weight Loss Among Overweight Employees," *Journal of Occupational and Environmental Medicine* 49 (2007): 981.

CHAPTER 4: THAT NAGGING FEELING

72 **Russell's blocking for the Celtics:** Gilbert Rogin, "We Are Grown Men Playing a Child's Game," *Sports Illustrated* (Nov. 18, 1963).

73 **value ambiguity:** In *Predictably Irrational*, Dan Ariely describes telling a class of MIT students that he is planning to conduct three readings from Walt Whitman's *Leaves of Grass* the coming Friday evening. He asked half the students how much they would be willing to pay to listen, and he asked the other half how much they would have to be paid in order to go. Sure enough, students who were told that an Ariely recital was something you had to pay for were willing to bid to pay for it—on average, about two bucks—and were even willing to pay three bucks for a long reading. On the other hand, those who were asked how much they would have to be paid to listen to Ariely's reading demanded on average $2.70—and $4.80 to suffer through a long reading. See Dan Ariely, *Predictably Irrational* (2008): 42–43.

73 **How much to pay:** Uri Gneezy and Aldo Rustichini, "Pay Enough or Don't Pay at All," *Quarterly Journal of Economics* 115 (2000): 791.

75 **Anti-incentives to lie:** Louisa May Alcott, *Little Men: Life at Plumfield with Jo's Boys* (1871).

75 **Incentive training in the marines:** www.marines.cc/content/view/32/33/.

77 **Posner on stickK:** Eric Posner, "StickK Business," the University of Chicago Faculty Blog, Dec. 4, 2007, http://uchicagolaw.typepad.com/faculty/2007/12/stickk-business.html.

78 **My book on gay rights:** Ian Ayres and Jennifer Gerarda Brown, *Straightforward: How to Mobilize Heterosexual Support for Gay Rights* (2005).

78 **The Restrained Radical's donation to NARAL:** "I Donated to NARAL," Restrained Radical, Sept. 19, 2008, http://restrainedradical.blogspot.com/2008/09/i-donated-to-naral.html.

79 **Cialdini's work on influencing people:** Robert B. Cialdini, *Influence: The Psychology of Persuasion* (1998); Noah J. Goldstein, Steve J. Martin, and Robert B. Cialdini, *Yes! 50 Scientifically Proven Ways to Be Persuasive* (2008); and Robert B. Cialdini, "Crafting Normative Messages to Protect the Environment," *Current Directions in Psychological Science* 12 (2003): 105.

80 **Hotel towels:** Daniel Finkelstein, "The Persuaders: Robert B. Cialdini and the Science of Persuasion," *The Times* (London), Nov. 5, 2007, http://entertainment.timesonline.co.uk/tol/arts_and_entertainment/books/book_extracts/article2804923.ece.

80 **Graduate student lunches:** Stephanie Tang, "No Such Thing as a Free Lunch: Peer Comparison, Normative Messages, and Self-Presentation," unpublished manuscript.

82 **Peer pressure to reduce energy usage:** Ian Ayres and Barry Nalebuff, "Peer Pressure," *Forbes,* Apr. 11, 2005; Ian Ayres, Sophie Raseman, and Alice Shih, "Evidence from Two Large Field Experiments That Peer Comparison Feedback Can Reduce Residential Energy Usage" (July 16, 2009), available at http://papers.ssrn.com/sol3/papers.cfm?abstract_id=1434950. Our work builds directly on Cialdini's own pilot study in San Marcos, California, where researchers placed door hangers with comparative energy usage on the front doorknobs of 290 homes; see P. Wesley Schultz, Jessica M. Nolan, Robert B. Cialdini, Noah J. Goldstein, and Vladas Griskevicius, "The Constructive, Destructive, and Reconstructive Power of Social Norms," *Psychological Science* 18 (2007): 429.

84 **Sexual orientation discrimination:** "Fortune 500," EqualityForum, www.equalityforum.com/fortune500/.

87 **Self-blackmail as commitment:** Thomas C. Schelling, *Strategies of Commitment and Other Essays* (2006): 79. The *Primetime* story is taken from Avinash Dixit and Barry Nalebuff, *The Art of Strategy* (2008), 14–17.

89 **Wolfers running the Stockholm Marathon:** Justin Wolfers, "Credible Commitments and Embarrassment, or Why I'm Telling You I'm Running the Stockholm Marathon," Feb. 27, 2008, http://freakonomics.blogs .nytimes.com/2008/02/27/credible-commitments-and-embarrassment- or-why-im-telling-you-im-running-the-stockholm-marathon/.

90 **Naked in Tauranga:** Lauren Owens, "Smoking Quitter Wins Bet, Mate Gets Naked, *New Zealand Herald,* May 30, 2008.

91 **The virtual wife:** Metaboinfo.com, www.metaboinfo.com/okusama/ (last visited on July 12, 2009); Rebecca Milner, "Virtual Wife Reminds You to Eat Healthy," CScout Japan, Oct. 21, 2008, www.cscoutjapan.com/en/ index.php/virtual-wife-reminds-you-to-eat-healthy/; Freakonomics blog, "The Virtual Nagging Wife Diet," Oct. 21, 2008, http://freakonomics .blogs.nytimes.com/2008/10/31/the-virtual-nagging-wife-diet/.

93 **Mazar's honesty study:** Nina Mazar and Dan Ariely, "Dishonesty in Everyday Life and Its Policy Implications," *Journal of Public Policy* 25 (2006): 1; Nina Mazar, On Amir, and Dan Ariely, "The Dishonesty of Honest People: A Theory of Self-Concept Maintenance," *Journal of Marketing Research* 45 (2008): 633–44; and Dan Ariely, *Predictably Irrational: The Hidden Forces That Shape Our Decisions* (2008): 218–22.

93 **Reagan in Russian:** www.reagan.utexas.edu/archives/speeches/ 1987/120887c.htm.

96 *Knocked Up*'s **commitment contract:** "Knocked Up Beard Joke," www.soundboard.com/sb/Knocked_up_Beard_Joke.aspx*Soundboard* (last visited on July 12, 2009).

97 **Fatbet cashless motivators:** http://fatbet.net/aboutRules.aspx.

98 **Group weight-loss contracts:** Robert W. Jeffery, Wendy M. Gerber, Barbara S. Rosenthal, and Ruth A. Lindquist, "Monetary Contracts in Weight Control: Effectiveness of Group and Individual Contracts of Varying Size," *Journal of Consulting and Clinical Psychology* 51 (1983): 242.

CHAPTER 5: MAINTENANCE AND MINDFULNESS

100 **Ryan Benson's failure:** " 'Biggest Loser: Where Are They Now?," *Today*, Dec. 16, 2008, www.msnbc.msn.com/id/28239000?pg=3#TDY _BLoser_WATN2; and Ryan Benson, "The Biggest Loser Questions and Answers . . ." Ryan C. Benson's MySpace Blog, Feb. 14, 2007, http://blogs .myspace.com/index.cfm?fuseaction=blog.view&friendID=100411667&- blogID=229723246.

102 **Andy's contract:** "STLSpidey's Profile," stickK, http://www.stickk.com/ members/commitment.php/cid/30237 (last visited on Aug. 22, 2009).

104 **Weight Watchers study:** Stanley Heshka et al., "Weight Loss with Self- help Compared with a Structured Commercial Program: A Randomized Trial," *JAMA* 289 (2003): 1792, available at http://jama.ama-assn.org/ cgi/content/full/289/14/1792. See also the discussion in the afterword of my book *Super Crunchers*.

104 **Preventing weight regain:** Christina Garcia Ulen, Mary Margaret Huizinga, Bettina Beech, and Tom A. Elasy, "Weight Regain Prevention," *Clinical Diabetes* 26 (2008): 100; see also Laura P. Svetkey et al., "Comparison of Strategies for Sustaining Weight Loss: The Weight Loss Maintenance Randomized Controlled Trial," *JAMA* 299 (2008): 1139.

105 **Counting calories:** See, for example, Mary L. Klem et al., "A Descriptive Study of Individuals Successful at Long-Term Maintenance of Substantial Weight Loss, *American Journal of Clinical Nutrition* 66 (1997): 239–46; Thomas C. Schelling, *Strategies of Commitment and Other Essays* (2006), 40 (distinguishing committing to effort from committing to results).

105 **Wing's work:** Rena R. Wing, Deborah F. Tate, Amy A. Gorin, Hollie A. Raynor and Joseph L. Fava, "A Self-Regulation Program for Maintenance of Weight Loss," *New England Journal of Medicine* 355 (2006): 1563; Kelly S. Dale et al., "Determining Optimal Approaches for Weight Maintenance: A Randomized Controlled Trial," *CMAJ* 180 (2009): E39; Brown University Media Relations, "Daily Weighing and Quick Action Keeps Pounds Off, Study Shows," Oct. 11, 2006, www.brown.edu/Administration/ News_Bureau/2006–07/06–035.html; Nanci Hellmich, "Weight War Can Be Never-Ending," *USA Today*, Oct. 16, 2005; M. T. McGuire, Rena R. Wing, and J. Hill, "The Prevalence of Weight Loss Maintenance Among American Adults," *International Journal of Obesity* 23 (1999): 1314; M. L. Klem, Rena R. Wing, M. T. McGuire, H. M. Seagle, and J. O. Hill, "A Descriptive Study of Individuals Successful at Long-Term Maintenance of Substantial Weight Loss," *American Journal of Clinical Nutrition* 66 (1997): 239; M. L. Butryn, S. Phelan, J. O. Hill, and Rena R. Wing, "Consistent Self Monitoring of Weight: A Key Component of Successful

Weightloss Maintenance" *Obesity* 15 (2007): 3091; G. C. Ulen, M. M. Huizinga, B. Beech, and T. A. Elasy, "Weight Regain Prevention," *Clinical Diabetes* 26 (2008): 100; E. C. Weiss, D. A. Galuska, L. Kettel Khan, C. Gillespie, and M. K. Serdula, "Weight Regain in U.S. Adults Who Experienced Substantial Weight Loss, 1999–2002," *American Journal of Preventative Medicine* 33 (2007): 34; M. R. Lowe, K. Miller-Kovach, and S. Phelan, "Weight-Loss Maintenance in Overweight Individuals One to Five Years Following Successful Completion of a Commercial Weight-Loss Program," *International Journal of Obesity and Related Metabolic Disorders* 25 (2001): 325; and T. A. Wadden and D. L. Frey, "A Multicenter Evaluation of a Proprietary Weight Loss Program for the Treatment of Marked Obesity: A Five-Year Follow-up," *International Journal of Eating Disorders* 22 (1997): 203.

106 **"Like"-speak:** You can hear "like"-speak in action in this great piece of slam poetry: "Like Lilli Like Wilson," www.youtube.com/watch?v= Azu8XWcHzFM.

107 **Stepping on the scale:** R. Baker and D. Kirschenbaum, "Self-Monitoring May Be Necessary for Successful Weight Control," *Behavior Therapy* 24 (1993): 377; K. Boutelle and D. Kirschenbaum, "Further Support for Consistent Self-Monitoring as a Vital Component of Successful Weight Control," *Obesity Research* 6 (1998): 219; R. A. Carels, L. A. Darby, S. Rydin, O. M. Douglas, H. M. Cacciapaglia, and W. H. O'Brien, "The Relationship Between Self-Monitoring, Outcome Expectancies, Difficulties with Eating and Exercise, and Physical Activity and Weight Loss Treatment Outcomes," *Annals of Behavioral Medicine* 30 (2005): 182; J. A. Linde, R. W. Jeffery, S. A. French, N. P. Pronk, and R. G. Boyle, "Self-Weighing in Weight Gain Prevention and Weight Loss Trials," *Annals of Behavioral Medicine* 30 (2005): 210; Linda Gonder-Frederick, "Self-Monitoring in Lifestyle Change Programs," *Staff Update, University of Virginia Health Systems,* Fall 2006, www.healthsystem.virginia.edu/ internet/ican/fall06.cfm; and Rena R. Wing, Deborah F. Tate, Amy A. Gorin, Hollie A. Raynor, and Joseph L. Fava, Statement Accompanying "A Self-Regulation Program for Maintenance of Weight Loss," *New England Journal of Medicine* 355 (2006): 1563.

107 **Not stepping on the scale:** "Biggest Loser—Where Are They Now?" Fit Cysters, Dec. 19, 2008, www.fitcysters.com/group/biggest_loser/forum/ topics/biggest-loser-where-are-they.

108 **Wing's randomized weight-maintenance study:** Rena R. Wing, Deborah F. Tate, Amy A. Gorin, Hollie A. Raynor, and Joseph L. Fava, "A Self-Regulation Program for Maintenance of Weight Loss," *New England Journal of Medicine* 355 (2006): 1563.

109 **How much weight dieters want to lose:** Gary D. Foster, Thomas A. Wadden, Renee A. Vogt, and Gail Brewer, "What Is a Reasonable Weight Loss? Patients' Expectations and Evaluations of Obesity Treatment

Outcomes," *Journal of Consulting and Clinical Psychology* 65 (1997): 79; and Weight Watchers, "Scientific Rationale for the 10% Difference," press release, www.weightwatchers.com/about/prs/ wwi_template.aspx?GCMSID=1000701 (last visited on August 21, 2009).

110 **Goal-setting advice:** NIH, Guidelines on Overweight and Obesity: Electronic Textbook, www.nhlbi.nih.gov/guidelines/obesity/e_txtbk/ txgd/4311.htm.

110 **Proofreading experiment:** Daniel Ariely and Klaus Wertenbroch, "Procrastination, Deadlines and Performance: Self-Control by Precommitment," *Psychological Science* 13 (2002): 219. The postmodern-text generator is available at www.elsewhere.org/pomo/.

111 **Spending on vices:** Klaus Wertenbroch, "Consumption Self-Control by Rationing Purchase Quantities of Virtue and Vice," *Marketing Science* 17 (1998): 317.

114 **Coffee cards:** Ran Kivetz, Oleg Urminsky, and Yuhuang Zheng, "The Goal-Gradient Hypothesis Resurrected: Purchase Acceleration, Illusionary Goal Progress, and Customer Retention," *Journal of Marketing Research* 43 (2006): 39.

115 **Compassion Korea:** Minjung Koo and Ayelet Fishbach, "Dynamics of Self-Regulation: How (Un)accomplished Goal Actions Affect Motivation," *Journal of Personality and Social Psychology* 94 (2008): 183.

117 **Flexible commitments:** K. C. Kirby, D. B. Marlowe, D. S. Festinger, R. J. Lamb, and J. J. Platt, "Schedule of Voucher Delivery Influences Initiation of Cocaine Abstinence," *Journal of Consulting and Clinical Psychology* 66 (1998): 761; J. Roll, S. T. Higgins, G. J. Badger, "An Experimental Comparison of Three Different Schedules of Reinforcement of Drug Abstinence Using Cigarette Smoking as an Exemplar," *Journal of Applied Behavioral Analysis* 29 (1996): 495; and Barry Nalebuff and Ian Ayres, *Why Not? How to Use Everyday Ingenuity to Solve Problems Big and Small* (2003): 21.

117 **Flexible mortgages:** "Holidays of a Lifetime," www.yourmortgage.co.uk/ 07_flex/holiday.htm; "Mortgage Payment Holidays Rise in Popularity," Mar. 15, 2007, www.home.co.uk/guides/news/tmc.htm?10735; and "Latest Property and Finance News: Two Million Mortgagees Are Considering Taking a Mortgage Payment Holiday in 2009," http://firstrung.co.uk/ articles.asp?pageid=NEWS&articlekey=10785&cat=44–0-0 (last visited on Aug. 22, 2009).

118 **The effectiveness of specific goals:** E. A. Locke and G. P. Latham, *A Theory of Goal Setting and Task Performance* (1990).

119 **Skinner's pigeon experiments:** Gail B. Peterson, "A Day of Great Illumination: B. F. Skinner's Discovery of Shaping," *Journal of Experimental Analysis of Behavior* 82 (2004): 317.

CHAPTER 6: WHAT COMMITMENTS SAY ABOUT YOU

122 **Brief talk:** Ian Ayres, "I'll Be Brief," Freakonomics blog, May 19, 2008, http://freakonomics.blogs.nytimes.com/2008/05/19/ill-be-brief/.

123 **Woody Allen's therapist:** Glen O. Gabbard and Krin Gabbard, *Psychiatry and the Cinema* (1999): 124.

124 **Refunds from easyJet:** Miles Brignall, "Triple Refund? It's Not as Easy as They Say," *Guardian,* March 8, 2008, www.guardian.co.uk/money/2008/mar/08/consumeraffairs.

126 **Approving mergers:** Ian Ayres and Stephen F. Ross, " 'Pro-competitive Executive Compensation' as a Condition for Approval of Mergers That Simultaneously Exploit Consumers and Enhance Efficiency," *Canadian Competition Record* 19 (1998): 18.

127 **BeniComp:** "BeniComp Advantage Video Highlights," www.benicompadvantage.com/products/video_highlights.htm (last visited on August 27, 2009); Victoria E. Knight, "Wellness Programs May Face Legal Tests; Plans That Penalize Unhealthy Workers Could Get Tighter Rules," *Wall Street Journal,* Jan. 16, 2008; Beth Baker, "Now, the Stick: Workers Pay for Poor Health Habits," *Washington Post,* Nov. 13, 2007, HE01; Robert Pear, "Congress Plans Incentives for Healthy Habits," *New York Times,* May 10, 2009, A16; "Big Stink: Employers Are Divided on Fines for Smokers," *Workforce Management,* Apr. 28, 2008, www.workforce.com/section/00/article/25/50/01.html; "stickK Corporate and Institutional Solutions," www.stickk.com/corporate.php (last visited on August 27, 2009); and U.S. Senate Republican Policy Committee, "Federal Constraints on Healthy Behavior and Wellness Programs: The Missing Link in Health Care Reform," Apr. 21, 2009, available at http://rpc.senate.gov/public/_files/042109FederalConstrantsonHealthyBehaviorandWellness.pdf. See also CBS News with Dean Reynolds (Nov. 6, 2007), www.cbsnews.com/video/watch/?id=3462888n&tag=related;photovideo.

130 **"Reform Legislation":** Steven A. Burd, "How Safeway Is Cutting Health-Care Costs," *Wall Street Journal,* June 12, 2009; Congressional Budget Office, "Health Care and Behavioral Economics," May 29, 2008, www.cbo.gov/ftpdocs/93xx/doc9317/05–29-NASI_Speech.pdf; "What Makes Orszag Run?" *New Yorker,* Apr. 29, 2009, www.newyorker.com/online/blogs/newsdesk/2009/04/george-w-bush-library-orszag-anticharity.html; and Catherine Rampell, "Do It or Pay," Economix blog, Apr. 10, 2009, http://economixblogs.nytimes.com/2009/04/10do-it-or-pay/.

131 **Preventing employment discrimination:** Ian Ayres and Jennifer Gerarda Brown, "Privatizing Gay Rights with Non-discrimination Promises Instead of Policies," *Economists' Voice* 2 (2005): art. 11.

134 **Improving the First Amendment:** Ian Ayres, "Compensation for Reckless Reporting 2," Balkinization, Mar. 2, 2005, http://balkin.blogspot.com/2005_02_27_balkin_archive.html; Ian Ayres, "Compensating for Reckless

Reporting," Balkinization, Feb. 21, 2005, http://balkin.blogspot.com/ 2005/02/compensating-for-reckless-reporting.html; Ian Ayres, First Amendment Bargains, *Yale Journal of Law and the Humanities* 18 (2006): 178, and *New York Times Co. v. Sullivan*, 376 U.S. 254 (1964).

136 **Voting commitment contracts:** Ian Ayres, "A Political 'Do Not Call' List," Freakonomics blog, Oct. 22, 2008, http://freakonomics.blogs.nytimes.com/ 2008/10/22/a-political-do-not-call-list/; and Dean Karlan, "How to Make a Decision and then Stick to It," *Financial Times*, Oct. 21, 2008.

136 **Prohibitions on paying voters:** California Election Code, Section 18521(a) (2000); and Alan S. Gerber and Donald P. Green, "The Effects of Canvassing, Telephone Calls, and Direct Mail on Voter Turnout: A Field Experiment," *American Political Science Review* 94 (2000): 653.

138 **Committing to conservation:** Ian Ayres, "A Commitment Device for Energy Conservation," Freakonomics blog, Nov. 14, 2008, http:// freakonomics.blogs.nytimes.com/2008/11/14/a-commitment-device-for- energy-conservation/; Ian Ayres and Barry Nalebuff, "Your Personal Climate Exchange," *Forbes*, Nov. 24, 2008; Ian Ayres, "Want a Politically Viable Gas Tax? Make It Voluntary," Freakonomics blog, Mar. 10, 2009, http://freakonomics.blogs.nytimes.com/2009/03/10/want-a-politically- viable-gas-tax-make-it-voluntary/; Ian Ayres and Barry Nalebuff, "A Voluntary Gas Tax," *Forbes*, March 16, 2009; and "Chicago Climate Exchange," Wikipedia, http://en.wikipedia.org/wiki/Chicago_Climate _Exchange (last visited on August 27, 2009).

142 **Do not pay contracts:** Illinois Gaming Board, "Statewide Voluntary Self- Exclusion Program for Problem Gamblers," www.igb.state.il.us/ selfxclude/ (last visited on August 27, 2009); 86 Illinois Administrative Code, Section 3000.700–790; Lee Fennell, "Slices and Lumps," 2008 Coarse Lecture on Law and Economics, Feb. 19, 2008, www.law .uchicago.edu/node/1108; Ted Gregory, "What Happens at Elgin Casino Stays at Elgin Casino, Man Finds," *Chicago Tribune*, Aug. 17, 2007, p. 1; Harrah's Entertainment, "Responsible Gaming," www.harrahs .com/harrahs-corporate/about-us-responsible-gaming.html (last visited on August 27, 2009); Jay Bhattacharya and Darius Lakdawalla, "Time- Inconsistency and Welfare," National Bureau of Economic Research, Working Paper No. 10345 (2004), available at www.nber.org/ papers/w10345; "Illinois Lottery Self-Exclusion from Play and Prize Payment Agreement," available at www.illinoislottery.com/ Selfxclude/SelfExclusionForm.pdf; and Illinois Gaming Board, "Gaming Board Strengthens Self-Exclusion Program," press release, available at www.igb.state.il.us/whatsnew/sepchange060622.pdf.

CHAPTER 7: ANTONIO'S PROBLEMS

146 **Agreeing to arm-breaking contracts:** Samuel A. Rea Jr., "Arm-Breaking, Consumer Credit and Personal Bankruptcy," *Economic Inquiry* 22 (1984):

188; Aristedes N. Hatzis, "Having the Cake and Eating It Too: Efficient Penalty Clauses in Common and Civil Contract Law," *International Review of Law and Economics* 22 (2002): 381; E. Allan Farnsworth, *Farnsworth on Contracts* (1999): 787; Harry Potter Wiki, "Unbreakable Vow," http://harrypotter.wikia.com/wiki/Unbreakable_Vow (last visited on August 28, 2009).

150 **Foreclosure-related suicides:** Dan Childs, "Foreclosure-Related Suicide: Sign of the Times," *ABC News,* July 25, 2008, http://abcnews.go.com/Health/DepressionNews/story?id=5444573&page=1.

150 **Schilling's incentives:** Gordon Edes, "Sources: Schilling Out Until at Least All-Star Break," *Boston Globe,* Feb. 8, 2008, p. E1; Sarah Green, "Curt Schilling: Staying in Boston and Staying Fit," Umpbump, Nov. 6, 2007, http://umpbump.com/press/2007/11/06/curt-schilling-staying-in-boston-and-staying-fit/; Curt Schilling, "Done," Curt Schilling's 38 Pitches, Nov. 6, 2007, http://38pitches.weei.com/sports/boston/baseball/curt-schilling/general/2007/11/06/done/; James Surowiecki, "The Fatal-Flaw Myth," *New Yorker,* July 31, 2006; E. E. Jones and Richard E. Nisbett, "The Actor and the Observer: Divergent Perceptions of the Causes of Behavior," in *Attribution: Perceiving the Causes of Behavior,* ed. E. E. Jones et al. (1972): 79; Scott T. Allison and David M. Messick, "The Group Attribution Error," *Journal of Experimental Social Psychology* 21 (1985): 563; Miles Hewstone, "The 'Ultimate Attribution Error'? A Review of the Literature on Intergroup Causal Attribution," *European Journal of Social Psychology* 20 (1990): 311; Jon Hanson and David Yosifon, "The Situation: An Introduction to the Situational Character, Critical Realism, Power Economics, and Deep Capture," *University of Pennsylvania Law Review* 152 (2003): 129.

152 **Torre contract offer:** Barry Schwartz, "Bonus Babies," *New York Times,* Oct. 24, 2007.

153 **Franklin's quest for perfection:** Benjamin Franklin, *The Autobiography of Benjamin Franklin* (1868; rpt., 1950).

153 **Addiction transfer:** Jane Spencer, "The New Science of Addiction: Alcoholism in People Who Had Weight-Loss Surgery Offers Clues to Roots of Dependency," *Wall Street Journal,* July 18, 2006; Maryls Johnson, *Cross-Addiction: The Hidden Risk of Multiple Addictions* (1999); S. Sussman and D. S. Black, "Substitute Addiction: A Concern for Researchers and Practitioners," *Journal of Drug Education* 38 (2008): 167.

154 **James Joyce's refrigerator:** Cartoon by David Jacobson, published in *The New Yorker,* September 25, 1989.

156 **Obama's smoking habit:** Jake Tapper, "President Obama Invokes Own Struggle with Cigarettes as He Signs Tobacco Bill," *ABC News,* June 22, 2009, http://blogs.abcnews.com/politicalpunch/2009/06/president-obama-invoked-own-struggle-with-cigarettes-as-he-signs-tobacco-

bill.html; and Kenneth R. Bazinet, "President Obama Not Perfect: Former Smoker Admits to Occasional Cigarette," *New York Daily News,* June 23, 2009.

156 **Ego depletion:** Roy E Baumeister, Ellen Bratslavsky, Mark Muraven, and Dianne M. Tice, "Ego Depletion: Is the Active Self a Limited Resource?," *Journal of Personality and Social Psychology* 74 (1998): 1252; Matthew T. Gailliot, Roy F. Baumeister, C. Nathan DeWall, Jon K. Maner, E. Ashby Plant, Dianne M. Tice, and Lauren E. Brewer, "Self-Control Relies on Glucose as a Limited Energy Source: Willpower Is More Than a Metaphor," *Journal of Personality and Social Psychology* 92 (2007): 325; Nicole L. Mead, Roy F. Baumeister, Francesca Gino, Maurice E. Schweitzer, and Dan Ariely, "Too Tired to Tell the Truth: Self-Control Resource Depletion and Dishonesty," *Journal of Experimental Social Psychology* 45 (2009): 594; Roy F. Baumeister, Matthew Gailliot, Nathan DeWall, and Megan Oaten, "Self-Regulation and Personality: How Interventions Increase Regulatory Success, and How Depletion Moderates the Effects of Traits on Behavior," *Journal of Personality* 74 (2006): 1773; Matthew T. Gailliot, E. Ashby Plant, David A. Butz, and Roy F. Baumeister, "Increasing Self-Regulatory Strength Can Reduce the Depleting Effect of Suppressing Stereotypes," *Personality and Social Psychology Bulletin* 33 (2007): 281; Sandra Aamodt and Sam Wang, "Tighten Your Belt, Strengthen Your Mind," *New York Times,* Apr. 2, 2008.

161 **Ego depletion and discrimination:** Matthew T. Gailliot, Roy F. Baumeister, C. Nathan DeWall, Jon K. Maner, E. Ashby Plant, Dianne M. Tice, and Lauren E. Brewer, "Self-Control Relies on Glucose as a Limited Energy Source: Willpower Is More Than a Metaphor," *Journal of Personality and Social Psychology* 92 (2007): 325; Matthew T. Gailliot, B. Michelle Peruche, E. Ashby Plant, and Roy F. Baumeister, "Stereotypes and Prejudice in the Blood: Sucrose Drinks Reduce Prejudice and Stereotyping," *Journal of Experimental Social Psychology* 45 (2009): 288; and Matthew T. Gailliot, E. Ashby Plant, David A. Butz, and Roy F. Baumeister, "Increasing Self-Regulatory Strength Can Reduce the Depleting Effect of Suppressing Stereotypes," *Personality and Social Psychology Bulletin* 33 (2007): 281.

163 **Hedonic treadmill:** Philip Brickman and Donald Campbell, "Hedonic Relativism and Planning the Good Society," in *Adaptation-Level Theory: A Symposium,* ed. M. H. Appley (1971): 287–302.

164 **Punished by rewards:** Alfie Kohn, *Punished by Rewards* (1993). Daniel Pink also identifies seven reasons incentives may backfire. Daniel H. Pink, *Drive* (2009).

164 **Maximizers and satisficers:** Sheena S. Iyengar, R. A. Wells, and B. Schwarz, "Doing Better but Feeling Worse: Looking for the 'Best' Job Undermines Satisfaction," *Psychological Science* 17 (2005): 143–50.

CHAPTER 8: A COMMITMENT STORE

167 **Dean's accomplishments:** David Leonhardt, "The Future of Economics Is Not So Dismal," *New York Times,* Jan. 10, 2007; Steven D. Levitt, "Congratulations to Dean Karlan," Freakonomics blog, Dec. 5, 2007, freakonomics.blogs.nytimes.com/2007/12/05/congratulations-to-dean-karlan/; "Microfinance and Commitment Contracts," Dean Karlan, interviewed by Romesh Vaitilingam, Vox podcast, July 31, 2009, www.voxeu.eu/index.php?q ode/3826; "Stick with It" *Brian Lehrer Show,* May 4, 2009, www.methings.com/podshows/4678188.

169 **Dean's work with commitments:** Nava Ashraf, Dean S. Karlan, and Wesley Yin, "Tying Odysseus to the Mast: Evidence from a Commitment Savings Product in the Philippines," *Quarterly Journal of Economics* 121 (2006): 635; "Put Your Money Where Your Butt Is: A Commitment Contract for Smoking Cessation," Xavier Giné, Dean Karlan, and Jonathan Zinman, World Bank Policy Research Working Paper No. 4,985 (July 3, 2009); "Microfinance and Commitment Contracts," Dean Karlan, interviewed by Fomesh Vaitilivgam, Vox podcast, July 31, 2009, www.voxeu.org/index.php?q ode/3826; and Steven D. Levitt, "Put Your Money Where Your Butt Is," Freakonomics blog, Mar. 10, 2008, http://freakonomics.blogs.nytimes.com/2008/03/10/put-your-money-where-your-butt-is/.

173 **Barry and my *Forbes* column:** Ian Ayres and Barry Nalebuff, "Skin in the Game," *Forbes,* Nov. 13, 2006.

177 **Transaction costs:** stickK currently keeps a percentage ("plus any electronic processing fees incurred by stickK") of forfeited stakes to compensate for transaction and accounting costs. See www .stickk.com/faq/tac/.

178 **stickK in the news:** www.stickk.com/about.php#news.

179 **Posner's critique:** Eric Posner, "StickK Business," University of Chicago Faculty Blog, Dec. 4, 2007, http://uchicagolaw.typepad.com/faculty/2007/12/stickk-business.html.

179 **unstickK.com:** Joshua Gans, "Business Killing Idea," Core Economics, Nov. 17, 2007, http://economics.com.au/?p=1198.

180 **stickK spin-offs:** "Tyler and I Have a New Business," Marginal Revolution, Feb. 27, 2008, www.marginalrevolution.com/marginalrevolution/2008/02/tyler-and-i-hav.html.

184 **A failed stickK commitment:** Haley Cohen, "StuckK: A Website Cures Bad Habits—Sometimes," *New Journal* 41 (2008): 31 (a Yale University publication).

189 **zur Hausen's Nobel:** "The Nobel Prize in Physiology or Medicine 2008," *Nobelprize.org,* http://nobelprize.org/nobel_prizes/medicine/laureates/2008/index.html.

Index

ABOUT THE AUTHOR

IAN AYRES is an economist and a lawyer. He is the William K. Townsend Professor at Yale Law School and a professor at Yale's School of Management. A columnist for *Forbes* magazine, Ayres is also a regular contributor to the *New York Times* Freakonomics blog. He has written ten other books, including *Super Crunchers,* which was a *New York Times* business bestseller and was named one of the Best Economics and Business Books of 2007 by *The Economist.* In 2006, he was elected to the American Academy of Arts and Sciences.